The Future of Religion and the Religion of the Future

Theodore John Rivers

UNIVERSITY PRESS OF AMERICA,® INC.
Lanham • Boulder • New York • Toronto • Plymouth, UK

University Press of America,® Inc.
4501 Forbes Boulevard
Suite 200
Lanham, Maryland 20706
UPA Acquisitions Department (301) 459-3366

Estover Road
Plymouth PL6 7PY
United Kingdom

Toward a Philosophy of Culture, Vol. 3

Printed in the United States of America
British Library Cataloging in Publication Information Available

Library of Congress Control Number: 2011934940
ISBN: 978-0-7618-5654-2 (clothbound : alk. paper)
eISBN: 978-0-7618-5655-9

∞™ The paper used in this publication meets the minimum requirements of American National Standard for Information Sciences—Permanence of Paper for Printed Library Materials, ANSI Z39.48-1992

Dedicated to
my wife, Doris Evans,
and to the memory of my brother,
William Gordon Rivers,
July 4, 1943–April 21, 2009

Contents

Preface

Some religions lose their importance and are superseded by others. Some lie hidden after an initial burst of activity to be reborn centuries later. Some persist in modified form, evolving from their original intent. Some transform both themselves and the world. Regardless where they originate, religions are expressive of human choices. Of those religions that have been lost and those that have survived, the future will introduce new varieties. Since religion is fundamental to human experience and since technology is a persistent and ever-present accompaniment to human reality, it is conceivable that these two activities might merge to create a new religious experience. We should ask not only what the future of religion might be, but we should also ask what the religion of the future could be. As this book indicates, the fundamental characteristics of technology as a new religion are already in existence, and have been in existence for quite some time. Except for a prophet, all manner of doctrines, rites, and rituals, even a sacred history, are already in place for technology. They are supported not only by technology's artifacts, but also by its procedures and methodologies.

It should be apparent that without technology, humanity would not have survived. And because technology is necessary for human survival, it is and has always been the means to salvation. It is irrelevant whether or not we acknowledge this truth, a truth that goes beyond any presumed importance of a voice thundering from the clouds, beyond the sermons of an iterant preacher, beyond a scribbler taking dictation from on high, beyond a thinker who devised an enumerated path to heaven, beyond an anthropomorphic imposition of the natural significance of the sun. Because we can speak of technology as the savior of humanity from many points of view, we can also speak of technology as a savior religiously. The religious relevancy of technology is derived from its non-religious potencies that are tied to technology's omnipresence and omnipotence.

If we were to examine the role of religion in society, we would encounter many interrelated features that facilitate its examination. In addition to social and economic, or political and cultural perspectives, it is the ontological perspective that reflects the most important point of view because it deals with the most fundamental underpinning of religion itself. Human reality analyzed from the ontological perspective manifests humanity's foundation. It reveals the essential characteristics of the being of human beings, characteristics that impact the very nature of humanity.

It is the intention of this book to evaluate and reveal the underlining truth of the being of technology in relationship to religion. Since technology may be vital to the emergence of new religious forms, it would be wise for us to take note of this relationship. Although there are believers in the present age who continue to pursue traditional religions, they seem to be unaware of the changes taking place all around them, of a type of transformation that the world has experienced many times before, in fact, every time a new religion has appeared. In that sense, our time is no different. What makes the present age unusual is simply stated: technology has reared its head as a predominant phenomenon because the past has set in place certain conditions that may transform our religious experiences. One of the most important conditions that qualifies as a characteristic of this transformation is Christianity because it has bestowed upon technology a forwardly directed intent.

Despite our arguments, it should be noted that we are not dealing with an inevitability, but with the result of our own choices. All religions are human creations that have appeared because of certain defining events. The present age relies upon certain historical and cultural events that hitherto did not exist. Consequently, new religions will reflect their place in time as previous religions reflected theirs.

Acknowledgments

Some chapters have been previously published. Chapter three incorporates what appeared as chapter seven, now modified, from my *Contra Technologiam: The Crisis of Value in a Technological Age*, Lanham: University Press of America, 1993, pp. 83-92. Permission to use this material has been granted by the publisher, the University Press of America, a member of the Rowman & Littlefield Publishing Group. Chapter four is reprinted from *Technology in Society*, 28 (2006), 517-531, with slight modifications. Permission to use this material has been acknowledged by the publisher, Elsevier, Ltd.

Chapter One

Religion

An Attempt at a Definition

INTRODUCTION

Although it may be assumed that a word as common as religion would be easy to define, this assumption cannot be supported by its applications within Western cultures, and less so within non-Western cultures. Even in the Middle Ages, the word associated with religion was not *religio*, which is derived from classical Latin and is the root for the modern word religion, but *fides*, which meant faith in the Christian religion only. The word *religio* in medieval sources signified a motley of meanings, including piety, ascetic life, the rule of a religious order, and various related religious or ecclesiastical matters.[1] It was only in Western Europe during the Renaissance when *religio* was associated with the word we know as religion today.[2] Since that time, there have been many attempts to use and define this word, which might indicate, like the use of *religio* in the Middle Ages, that it may have no definite meaning, unless it is limited to a particular time and place. Because the circumstances in which religion is evident vary so widely, any attempt to seek a definition will be difficult, but these circumstances have not discouraged its pursuit. Regardless of its application, religion seems to be characterized by an inherent ambiguity that may be the cause of its different, and even contradictory definitions.

For example, religion has been defined as a belief in a divine mind and will that rules the universe and has a moral relationship with humanity (Martineau), but this definition is based on the presupposition that God or gods exist. It has been defined as the attitude humans give to the power(s) considered to have control over their interests and destinies (Pratt), but there is no evidence of this power. It has been defined as a feeling derived from an absolute dependence (Schleiermacher), but this definition is tied to an inherent inadequacy or insufficiency. It has been defined as the perfection of goodness that permeates humanity (Bradley), but goodness is an arbitrary abstraction and is culturally relative. It has been defined as an emotion based on a harmony between ourselves and the universe (McTaggart), but harmony is indefinite and simply means an agreement in feeling. It has been defined as a reaction of nature against human intelligence (Bergson), but this definition not only infers godlessness, it also is applicable to many other endeavors. It has been defined as a system of symbols that establishes powerful and long-lasting moods and motivations (Geertz), but this definition's inclusivity is the cause of its own ambiguity.[3]

The most obvious difference among these definitions concerns the presence or absence of a divine being, regardless whether it is thought to be supreme (God) or subordinate (a god). Although a definition may incorporate the existence of a divine being, how can human intelligence know of anything that transcends our ability to know? What is proposed as an answer to this question is an idea that allegedly bridges the gap between the reality that is known and a transcendence that is not. We must keep in mind that religious faith is not a matter of knowledge because faith is not an epistemological category. Unfortunately, religion has been appended to all manner of things, and has given them an aura of religiosity. Fortunately, religion has also been subjected to critical analysis that can expurgate its influence from these things.

Because faith is associated with an inner piety, the latter is closer to a more accurate meaning of what it represents rather than its outward manifestation.[4] It is a simple task to reify this inner piety, that is, to give an existence to an abstraction by making real a feeling. But unless some type of belief system accompanies an inner piety, it is not considered to be religion in the technical sense because religion is thought to require an organization in order to qualify as a genuine religion. Because a system has interrelated features or components, it may be perceived to have a type of organization. And since an organization is anything that is arranged into an ordered whole, a system may be included within this description. Although system and organization are not equivalent entities, they are related. If a set of beliefs has a systematic approach to religion, it may have some type of organization. Any systematic belief network would have greater influence than mere tradition. Belief systems, therefore, that presuppose some type of organization help to constitute a religion.

There are some religions that have a belief in divine beings, a morality appended to them, a conception of an alleged difference between sacred and profane objects, rituals that reflect this difference, even a worldview, but none of these features are essential characteristics of religion. The irony associated with religious belief as belief does not revolve around the influence of God or gods on humans, but the human perception of this influence.[5] And this perception revolves around ideas, not beings. To assume, for example, that Jesus Christ was divine came about because New Testament Christians were in awe of a carpenter's son who they thought was "the Son of God." The expectation of a savior for Jewish society, with which Jesus was associated, had been anticipated for a long time, and what came about had been set in place from the beginning. Likewise, the sun was thought to be divine by some people because its life-giving capability was considered to be a god, as the Nile river was considered to be the embodiment of a god because its waters gave birth to life-sustaining vegetation. A carpenter's son, the sun, and the Nile river were thought to be divine because their supporting societies promoted their divinity. As a result, they were social manifestations of religious beliefs.

Therefore, a major proposition associated with religion concerns whether it is necessarily tied to theism, that is, to the belief in God or gods. Certainly, this belief becomes the fulcrum around which many religions revolve, but the absence of a god as a feature of religion would not be an impossibility, although it may be uncommon. Because a god is not necessary for religion, as Buddhism or Jainism will attest, we must conclude that the presence of a god is not fundamental to its definition.[6] Likewise, the absence or denial of religious beliefs may prepare the way for new beliefs,[7] but there are characteristics that religion in general and all religions in particular seem to share, which include sociability, causality, practicality, rationality, and a belief system. We will have to determine if these characteristics are genuine, but we do know that they express themselves socially as well as ontologically. Hopefully, these characteristics will point the way to a definition.

Even if we disregard, or at least are mindful but do not support the controversy between idealism and realism, a definition may be attempted. Although religion may be theistic or atheistic, monotheistic or polytheistic, or even pantheistic, it is a human activity that is assumed to have a universal application. The word itself exemplifies a comprehensiveness that all societies allegedly share, whether they are prehistoric or historic, Eastern or Western, Christian or non-Christian, god-fearing or godless. To which it must be said that the idea of religion did not originate from totemism or shamanism, since both of them presuppose the existence of a religious feeling. Nor did it originate from magic, although there are vestiges of magic in many religions. Nor did it arise from a distinction between what is considered to be sacred and profane, nor from the notion of the soul. Religion did not arise from any of these ideas, although it is now tied to them, because a religious feeling would have to exist first before any religious manifestations or applications would be evident.

SOCIABILITY

As the word is understood, religion indicates social activity. It binds individuals to a group as it binds a group to individuals. The word religion (*religio*) is thought to be derived from the Latin *religare*, which means to bind or to tie, that is, religion concerns the effects of its prescriptions on a community of believers who are united by ideas that are held in common. Although the root of religion might be ancient, the word as it is used is modern. The importance of *religare* emphasizes the practical effects of religion.[8] Although practicality is another characteristic of religion, it is related to sociability. Without interaction among humans, religion would have a reduced effect. Indeed, a different type of religion would exist if only one person was involved or if there was no interaction with other members of a group, which for our purposes we may equate with society. Even the primitive beginning of religion in the Old Stone Age seems to indicate that spirits were worshipped for the good of the group, not the individual. If religion supports individuals, it does so through society, that is, individuals become important only as a consequence of society's presence. It is society by which individuals are recognized and from which they originate. The individual hermit living in the desert, for example, has no meaning to society if he has no influence on it, but he can acquire an influence if he inspires others to act likewise.

A god is a representation of its society, and societies are comprised of people. Yet a god is not society's embodiment, although its attributes display the qualities that a society holds most dear.[9] In effect, we might say that a god is human nature deified. What we know about a god is derived from self-reflection. A merciful god reveals the human ability of forgiveness as a vengeful god manifests the human inclination for retribution. Regardless of a god's qualities, they are embodiments of human qualities. Apart from the belief in miracles that defy nature, where would these qualities originate if not from us? Divinity reflects humanity, and humanity manifests divinity.

All gods are predicated on their attributes, but in order to be distinguishable as gods, they must be abstracted from the attributes that are given to them. Therefore, gods must be distanced from their attributes in order to substantiate their existence and to support the idea that indeed they are gods. But a god that has abstracted attributes—or abstracted anything—is a fiction. Since we often hear that we should look for the divine presence within us, this expression may be interpreted to mean that we should mimic a god's divine qualities that may encourage us to adopt its attributes. A more accurate way of stating this expression is that we should follow the human attributes associated with this god.[10] And when we unite our belief with the belief of others, we attain a conformity with their ethic and their corresponding values.

More than a god, the most important aspect of religion are the values it promotes, although values in themselves are not characteristics of religion. No irony is intended when it is said that although religion has values, they are not its characteristics. This relationship is true because values are effects, not causes, of religious belief. Characteristics are causes because they supposedly give rise to their effects, but a value is an effect, not a cause, because it is an adaptation of a point of view. A value is the subjectivization of the worth of its object, which may be either a physical entity or a mental construct, that is, a judgment must be in place first before anything can assume a value. Nevertheless, if all members of society manifest the same religious objectives, then its belief would be common to all, and its communality should uphold the values of that religion. All values are aided by the exercise of the will, but the world itself, with which they are associated, is valueless.

Sociability does not mean just friendly disposition. It does not mean sociality, which concerns living with others, but living amicably with them. Although there is a natural tendency to seek the companionship of other people, that is, to live in some type of social setting that demands the presence of others, sociability is far more elaborate than sociality because it concerns the integration of individual efforts into group activities. It means the character, quality, and ability of being sociable. Sociability presupposes a temperament for agreeable and social interaction that leads to cooperation, which is a shared, not an individual characteristic. Even children at a very early age develop a network of social values that includes the importance of cooperation.

We must ask what purpose would be served if believers had no common basis with others of their religion. What purpose would be served if believers did not share ideas, rituals, and beliefs that were held in common? Just as sociability fosters harmony because it promotes solidarity, any cooperation among members in society presupposes the existence of some type of social control. Sociability, like cooperation, is innate to humans. It concerns acting together toward shared goals that could be the achievement of some social activity, or simply the means of being together in order to further their identity. But we need to ask if religion deals with problems that are created by society that seek resolution, or by changes that occur within society, or both. If this question can be answered affirmatively, then religion becomes a mechanism for the maintenance and perpetuation of social norms, but it must be differentiated from piety. It becomes just one mechanism among many for society's approach to social control that also includes the state.

And yet our perception of a divine being's supernatural presence, or the feeling for a presumed realm beyond reality, is not derived from sense experience, but imposed on it. It a sense, we can say that its imposition infers an anticipation. Nevertheless, this perception or feeling may be the result of certain concepts or mental formulations that have also been analyzed by a long philosophical tradition.[11] It is not to be confused with the theory of innate ideas that has its roots in Platonism which claims that the foundation of all knowledge transcends sensory experience. If the validity of these concepts can be affirmed, then this phenomenon must be innate or *a priori*, that is, they must have originated as the result of cognitive changes that occurred in conjunction with human development. But the philosophical explanation of these concepts and their neurological (or physiological) explanation are two different, although perhaps overlapping interpretations. Any question that we ask of these concepts must precede questions of truth because truth is the result, not the cause, of conceptualization. For example, when Jesus Christ promoted himself as God, this idea expressed a conceptualization of divinity that his followers accepted.[12]

Because natural selection favored different types of human intelligence that concerns social behavior, it may be related to the development of the capability that processes data. Nevertheless, we do not know whether this mental development occurred gradually or suddenly after a threshold had been crossed. If mental concepts or formulations are inherited from the evolutionary past, then humans developed them, and they have become part of humanity's capability. We should be able to conclude, therefore, that concepts, including the concept of God or gods, are a human construction internalized.[13] We then externalize this concept when we project it to sacred objects, or self-proclaimed messiahs. Externalizations, such as burning bushes or divine prophets, are tied to their respective societies that ultimately are influenced by history and culture, that is, defined by free choice.

CAUSALITY

A cause is generally described as an agent or factor that produces an effect. In the modern sense, a cause is generally associated with the ancient Greek notion of efficient cause that describes the means or manner by which an effect is brought about. Because we attribute an affected difference with an effect, we associate change with the latter, and therefore, when we think of effect, we think of change. Nevertheless, a cause may not be necessary or sufficient for there to be an effect. A necessary and sufficient cause relates to an event that by itself alone will cause something else to occur. Because necessary and sufficient causes are rare, they are nearly impossible to confirm. Perhaps, they should be called "divine causes," but even here we lack the evidence to confirm them beyond a reasonable doubt. All that is really necessary for a condition to be a cause is to have a relationship with its effect, but this relationship does not have to be bound with its effect. Cause and effect do not have to be connected; they only have to be conjoined, that is, combined or joined together. The cause-effect relationship is not connective, but conjunctive, as the word "and" joins two sentences together.

We need not concern ourselves with the idea that change may not produce a universal effect, but only with the general idea that there is change in the world and that it fosters a condition that is different from, but influenced in some way by its cause, whatever the latter might be. Ultimately, causality concerns the creation of the universe and the question why there is something rather than nothing. Why are there planets, stars, and galaxies? What is the source of the energy of the universe, and why is there life? Although an effect may not necessarily follow from a cause, it may appear that it does. What we perceive to be ultimate causes do not have to be purposeful. In fact, there may be no effective relationship between cause and effect, but this presumed relationship does not prevent us from making a connection. To conclude that an event is the cause of another event because one precedes the other is to make a false conclusion, which in logic is known as the *post hoc, ergo propter hoc* ("after this, therefore because of this") fallacy. As we indicated above, events may be conjoined, but not connected.[14]

This argument is relevant to our discussion because all creation is considered to have a cause without which it would not exist. We are accustomed to expecting an effect to follow from a cause because a cause is presumed to be necessary in order to have an effect. This cause-effect relationship is based on a supposition that whenever a cause occurs, an effect will inevitably follow. But cause and effect may not occur simultaneously, and therefore, they may not be contiguous. The lack of contiguity may annul any direct and genuine relationship between them.

Regardless of the complexity associated with the questions asked above, an answer has been proposed whose agent is described as "God" or "god," which is the name given to the ultimate cause that is assumed to be essential for everything that follows. Unfortunately, this description eliminates the need for searching for a real answer, and it is a cause that shows the limitations of human knowledge. For an agent to be a cause, it must be assumed that it relates to an effect, that is, we must be able to understand why God or a god has the ability or the desire to bring change into the world, or why it is necessary to have any god at all. This knowledge is related to how we experience the world, that is, it is based on empirical knowledge, otherwise, it would lie beyond our ability to know. We should look for the source of this change, or any change, within the world itself without being diverted by phenomena that allegedly transcend it. Although the ancient world joined causality with teleology as the ultimate purpose of life, the modern world has abandoned the desire to do so.

The idea of power has been disassociated from arguments that deal with causality, but there is a connection between them in regard to the existence of a god. We assume that a god is powerful because it is presumed to have the ability to bring change into the world or can create things that did not exist before because it possesses powers that transcend the laws of nature. And this power is why we perceive this being to be a god. For example, it is said that Atlas was so strong he could hold up the world, that Jesus was able to change water into wine, that Isis was capable of bringing Osiris back to life every spring, that Yahweh could destroy a people provided the Jews obeyed him. And yet divine power is also thought to be imparted to humans, which is especially noteworthy in Judaism and Christianity, where the expression that humans are made in "the image of God" implies that the latter's power can be imitated and utilized by humans for their own benefit. This expression reveals more than a metaphoric representation. It reveals that humans have acquired divine attributes.[15]

But how do these arguments relate to religion? If they are based on assumptions, presuppositions, or errors, we may have to conclude that their claims are false. We associate gods with the ability to bring change into the world because of some inherent power, but are they actually agents of change? As we mentioned above, the lack of contiguity falsifies any claim to validity, as does the alleged conditionality of sufficiency and necessity. Causality is more likely to revolve around probability, not certainty of events. Although delineated as a characteristic of religion, causality is based on a false presupposition. It is assumed, not proven to be true. And, likewise, this assumption must be extended to the gods.

PRACTICALITY

Although causality is not a characteristic of religion, it is related to practicality. Because it is presupposed that religious ideas would be applied wherever possible, its application would lead to useful effects, that is, religion is not a theoretical, but a practical endeavor. It is immersed in the world, not disengaged from it, and because it is immersed in the world, it is possible to conceive of a god. Even when expressed otherworldly, religion is tied to the world. Despite its preoccupation with an alleged realm behind, above, and beyond the world, without the world, this alleged realm would not exist. We must observe the world first before we can possibly conceive of another. If the world is real, then all other worlds must be imaginary, but it is to the imaginary world that many people project their highest hopes because it is not subject to the rigors of reality. If we can be honest with ourselves, we should admit that we can know just the real world, which is the only world that can be understood, and it is understood empirically. Those who say that there is another realm—the abode of gods, spirits, and ancestors—pursue an illusion.

Because religion is practical, it may be assumed that its effects would have some social significance, that is, practicality reinforces the idea of sociability discussed above. Practicality revolves around the question of efficiency, which aims at the attainment of goals. This is to say that unless goals are attainable, they cannot be considered to be efficient. Therefore, religion promotes efficiency by stressing practicality. How much does religion's practicality help or hinder the quality of life? If it is helpful, how much does it manifest a useful function in human society? If religion is practical, it must be connected in some way to human volition because without the will, religion would have little lasting effect. Religion concerns the relationship of individual members of society, and it concerns an ethic that society strives to promote.

This relationship illustrates that religion is tied to morality because the ethic that individuals strive for is the morality that society has already set in place before the completion of its religion. Therefore, morals do not emanate from religion, but the reverse. A case in point is the action of theft, which did not originate as a sin punishable by religion, but a crime against society. Of course, if theft was universal, then everyone would be inconvenienced, and society would suffer. In order to prevent the latter, customs or laws have been established that prohibit theft and punish it as a crime. Whatever restrictions religion may have developed against theft are patterned on and are supplemental to society's prohibitions, but religion by itself does not have the ability to enforce morality. It is dependent on society to do this. We should ask ourselves how much morality is a reaction to the misgivings of the world, and how much it is based on irrationality or emotions such as fear or resentment.

Originally used by people to unify themselves, religion and ideas associated with it were later used to fight its own members who may have been resistant to its practices. Religion also enforces social utility, and utility revolves around survival. Thus, in a more religious age, or among more religious people, religion becomes repressive, and in a less religious age, or among less religious people, it becomes permissive. Although some religions are innately more repressive then others, all religions are restrictive to some extent. Because society cultivates the feeling that individuals are expendable, religion augments the idea that society should be favored whenever individuals conflict with it.

In its varied manifestations, religion reveals more than the imperfections of the world. It also reveals our own imperfections, as it also reveals our expectations. Although "the axial period," as noted by Karl Jaspers,[16] concerns a time in which humanity supposedly became aware of its ontological limitations, the solution to these limitations concerns the creation of religions in which powerful gods were endowed with qualities that far exceed human capability. As a result, societies that produced these gods were successful in subordinating humans to themselves on the one hand, while emphasizing individuality on the other. These societies established moral and social ideals for universal religions, and became far more practical. Previously, individuals were identified with their tribes, and their religions were tribal in nature,[17] but now they were encouraged to be self-reflective about the fulfillment of future goals. The axial period that developed over a vast expanse of time from the eighth to the second centuries BCE revealed humanity's inherent vulnerability that could be overcome once humans conformed to the paradigm their societies promoted. The axial period introduced an approach to reality that entailed conformity to an organized scheme of belief, but it also attacked the individuality it originally encouraged.

The underlying importance of the axial period concerns more than human imperfections because it also reveals a tendency for anthropomorphism. Humans create society in their own image. More than the gods fashioned from the wind and the rain, or the sun and the moon, the gods of the axial period are fashioned in the image of people. This understanding seems to agree with the idea that religion is a type of anthropomorphism.[18] If gods are seen as persons, then people may enter into personal relationships with them. Even if some of these gods, such as Yahweh or Brahman, were not thought to possess a human body, they were thought to be able to hear, speak, and see like humans, thereby eliminating problems of imagery. These gods lost their condition of non-being (because they lacked bodies) and acquired the quality of being. Ultimately, the axial period revealed the importance not so much that gods have become human, but ironically humans have become gods. The best example of these gods is Jesus Christ because he was proclaimed to be both a god and a man, a characteristic he shared in part with the rulers of ancient Mesopotamia and Egypt, who, although humans, were considered also to be gods. In Christianity, the power of the father, Yahweh, is transformed into his only begotten son, Jesus. Most emphatically we must say that anthropomorphism reflects our own self-importance because in the last resort, humans succeeded in supplanting the gods of nature with human gods. This process allowed humans to apply a practical matrix over nature, while it pacified their loss of naturalness.

RATIONALITY

Rationality or the capacity to use reason concerns reaching conclusions inductively or deductively, either by arriving at general principles based on individual facts or reaching conclusions that follow from premises. It is utilized for the attainment of goals that conform to the means applied because it is the result of the choices we make that are dependent on and are derivable from the world. Rationality helps us understand the world as the world influences our choices.

Because rationality is applied in the world, it signifies what is thought to be a practical application of this process, but rationality cannot eliminate emotion or emotional reactions to events. Rationality is a consequence of subjectivity which itself is grounded in our individual freedom to choose, but it can also be accompanied by a willingness to act. Rational contemplation is one thing, but rational application is another. Since we choose to act in certain ways, our choices indicate what we have deemed worthy of action. Religion, or at least an awareness that takes religious feeling into consideration, seems to be affected by such a choice, regardless of the cultural milieu, or the ideas needed to bring it about.

The attainment of goals, the freedom of choice and the willingness to act are integrated parts of a whole, but are they integrated parts of religion? Can the latter in its multifaceted applications exist without rationality? In general, we can say that rational conclusions are reached through the accumulation and assessment of evidence. But these conclusions do not eliminate emotion or irrational behavior, or reasoning without reference to facts or experience (*a priori*) that may have some influence on the formation of religion, or its modification, once religion has been established. People may act toward religious goals that are supported and promoted by the group, but these goals may not be affected by rational decisions.

By rationality, we do not mean rationalization that may stand in the way and oppose its application. Although rationalization may help religion on the one hand, it may distort it on the other. Religion is based on an interpretation that we make in the world. And because it is a worldly phenomenon, we render judgments from our experience. If the wind blows hard, for example, causing a tree to crash into a house killing the resident inside, it may be said that the gods were angry with the occupant and punished him. The wind, the tree, and the death that followed are interpreted from the circumstances, which help us to make connections between a presumed cause and its alleged effect. Depending on our interpretation, we can conclude that this death was or was not divinely inspired, that the resident had sinned or was an innocent victim of an accident, that nature was used by the gods to express their will or that the resident was at the wrong place at the wrong time. One interpretation makes a leap of faith and goes beyond reality, while the other limits itself to factual evidence alone and does not envision transcendent forces. This perception between understanding the world by itself and imposing some divine influence has a long tradition and goes back, at least, to the first Greek philosopher, Thales (seventh-sixth centuries BCE).

Based on the above, we might ask how much religion is the result of distortions of factual evidence. How much it is a misrepresentation of reality? If the world is irrational, then a rational response to it would be to act irrationally. But then, how would religion survive if some type of misrepresentation did not characterize it, if revelations or divine visitations or events contrary to nature did not intercede for reality? Indeed, what makes religion extraordinary is its exaggeration of the natural order, but the unnatural or ultrarational basis of religion seems to conflict with rationality. Even when we make rational choices, they may be modified or distorted. Religion is one such example, not only because it has a foundation in human evolution, but also because it is a social phenomenon that people share, and they may share in more things than the attributes of objective reality. Although a judgment may be rationally based, it may also be irrationally conceived. It does not seem likely that individual religions—primitive or modern—are the result of rational reflection, at least as far as the evidence indicates. There is every reason to conclude that although rationality may be assumed to be a characteristic of religion, the evidence refutes this assumption.

BELIEF SYSTEM

Religion is tied to belief, which is a condition that affords trust in something that is perceived to be true. But more than trust, belief is a feeling that brings to mind the definitions of religion by Schleiermacher and McTaggart described above. Because belief concerns the acceptance of certain tenets and the rejection of others, it is a type of conviction. It makes the believer confident in his ideas, and protects him from doubts that may question them. Nevertheless, all religions have beliefs, but they cannot sustain themselves and cannot foster sociability without some type of system that concerns the interaction of its individual elements into a whole.

More than the internal truths of a belief system, its external application is far more important. Although a belief system would vary dependent on the goals of its believers, any improvement in its application—which should extend its control—may be aligned with the demands of society, and should help to fulfill its calling. The perpetuation of a ritual or a doctrine, however subjectively applied, may be beneficial for the community. Indeed, religious beliefs need to be shared. They are made up of dogmas and creeds, myths and sacraments, rituals and institutions, and all manner of theologies. Although there may be no proof that these beliefs are true, they still need support from its believers to sustain them. Therefore, the greatest virtue of a religion is to believe in its belief, and the greatest misfortune is to disbelieve. When religion permeates society, indeed when any society becomes theocratic as in the European Middle Ages, religious practice may be perpetuated by extreme behavior: by imprisonment or the stake. Regardless of one's moral character, participation in a belief system erases all wrongs. Thus, in medieval Europe, monasticism became a way of life, although it distorted the sense of humanity's natural being with an obsession of an abnormality.

Belief and knowledge are tied together. It is possible to believe in something without having sufficient knowledge of it, but it is not possible to have knowledge without also having a belief about it. This is because knowledge implies an underlying belief where sufficient evidence exists that supports it. Because knowledge is based primarily on probabilities, it is based on the likelihood of something being true, not on the definitive conclusiveness or certainty that it is. Knowledge may be defined simply as true belief. For example, at one time it was common knowledge among many people that the earth was flat, and the available evidence indicated that this was true. Later, additional evidence was acquired that contradicted this knowledge and eventually replaced it with the knowledge that the earth is a sphere. Both the flatness and roundness of the earth were true, but at different times. At both times, there was a belief and a knowledge about the truth of the earth which indicated that not only has our knowledge of the earth changed, but also has our belief.

But belief in a divine being is different. It is not subject to the same criteria as factual truth. There may be no evidence supporting the existence of a god, or a reason for a ritual, but this does not prohibit its belief. Religion, like contextual inference, is based on an assumption in the absence of evidence. Although it is said that there may be much to believe about a god, little may be known. And when little is known, a mystery may be perpetuated. Knowledge accompanied by belief is based on subjectivity, but so is belief as belief sustainable solely in and by itself. Although a moving force, belief as belief is rarely altered by factual evidence alone. It needs factual evidence that is empirically sustainable by critical analysis in order to be verifiable. Nevertheless, subjective interpretation of the evidence that is used for knowledge is also used for the reinforcement of belief, that is, subjectivity becomes the basis both for factual evidence as the foundation of truth and the foundation of belief. Although contradictory, knowledge when supported by belief and belief when self-supporting have something in common; they are both subjectively grounded. Knowledge is dependent on a belief where sufficient evidence exists that supports it, but belief as belief is self-dependent.

Regardless of the difference between knowledge and belief, and regardless of the idea that something may be true even if no one believes or acknowledges it, people want a religion that they believe is true. In a general sense, we may say that in order to believe in anything, it must first seem to be true. A religion regarded to be false by its believers is a contradiction in terms.

Therefore, it is presumed true that all religions contain fundamental truths about a world beyond reality, or about the nature of God or gods, even those truths that are contradicted by other religions. If religions contain truths and these truths are contradicted by other truths, then religious truth is relative, but this is not what religions claim. If one truth may be substituted by another, then no truth is absolute. Religious belief is subject to human interpretation, and therefore, human subjectivity. It must be said in all honesty that religious belief still requires a leap of faith because it is possible to believe even when sufficient evidence is lacking. Because individual beliefs are interrelated, they constitute a system, however loosely it may be defined. The beliefs of eastern religions may not be as integrated and as well-organized as the beliefs of western religions, but their interrelatedness qualifies them as belief systems, nonetheless. We can conclude, therefore, that a belief system is a characteristic of religion.

CONCLUSION

Religion is an imprecise word. Although there have been many attempts to define this word, no attempt has been completely successful because religion's ambiguity lends itself to numerous interpretations and many implications. Supposedly, religion deals with forces that transcend the human ability to know or generate them, but the attempt to find a definition must apply to all religions regardless of time or place. This phenomenon, however, would reduce its meaning to the lowest common denominator.[19] Despite the importance of the characteristics of religion discussed above, all of them are subordinate to the notion of religious feeling. It seems likely that without this feeling, religion would not exist. Perhaps, religious feeling, or the ideas on which it is based, is the basis of religion. Nevertheless, for anyone to follow the traditions of one's family does not help in the least to understand the essence of religion; it merely perpetuates the belief system of a particular religion.

Although religious feeling is associated with religion, it has more to do with emotion. Regardless whether religion is the object of religious feeling, an emotive sensation may be its predominant quality. And this sensation is universally evident among all people, as it is influenced by free choice and the manifestations of individual societies. Since a feeling is an affective state, and an affective state is an emotion, the latter defies rationality. Although connected to sensation, feeling is subjective. It is a judgment of a consideration or deliberation, and is parallel to the idea that belief is a conviction as discussed above. Therefore, rationality is not a valid characteristic of religion, although falsely associated with it.

Despite the link between epilepsy (with its hallucinatory "revelations") and a heightened tendency toward religious feeling, the latter originates internally in the temporal lobe of the human brain before any external manifestation—a sacred rock or a messiah—is associated with it.[20] Therefore, religion results from the realization of certain conceptions. Mysticism, which attributes validity to a realm beyond perceptual or intellectual apprehension, such as that associated with Joan of Arc or Theresa of Avila, transforms its notion of reality into supernaturalism by creating an assortment of responses, characterized as apparitions and visions. But the truth remains: religion is not rational. Although religious feeling may originate internally, it is dependent on the world for its realization because the world is the recipient of its projection.

We have said that rationality is not a valid characteristic of religion, but by itself rationality is utilized for the attainment of goals that conform with the best means available, that is, rationality allows humans to live in the world as best they can—to act and react with the world's changes. Regardless how much religion is thought to be based on reason, individual religions present very irrational and illogical applications. There are even those who oppose reason and yet use it to defend their position that reason is useless in its opposition to transcendence or the mystical workings of the world, that reason is subordinate to faith or intuition because reason's opponents imagine another world to be superior to the real world. Religion may easily be transformed into ideology that can be used to unify people of a common belief.

Similarly, there may be no effective relationship between cause and effect in regard to religion, but we are not prevented from making a connection. Because we may wish to draw an inference between a presumed cause and its alleged effect, we configure the condition of one with that of the other. In addition to the arguments discussed above, we also attribute a feeling that may be the basis of religion to a cause that we assume comes from an alleged transcendent. But causality in itself has no validity. Just because a cause may exist does not mean necessarily that an effect must follow. We can assume, but may not confirm, if an effect is related to a cause, unless the effect has been immediately confirmed to be influenced by it. Unfortunately, such a process may be nearly impossible to determine. There may be causes in the world and there may be effects, but they may not be fundamentally related. Our ability to understand any relationship between them may be faulty, and likewise may also be our understanding that effects in the world are divinely inspired.

These arguments bring us to sociability, which as we said above concerns more than sociality or friendly disposition. Because sociability integrates individual efforts into group activities, its goal is cooperation, which is a shared, not an individual characteristic. Because it leads to agreeable and social interaction, sociability promotes a harmony that encourages a camaraderie among humans. The long-term effects of sociability concern the preservation of the group and the perpetuation of its values. This ability promotes innately, even instinctively the practices of religion. Therefore, sociability is a practical endeavor. As conceived, religion is tied to the world, which is parallel to this word's derivation from the Latin *religare*, which means to bind or to tie. This tendency enables religion to function as a mechanism of social control, which, because it is practical, fosters efficiency.[21] Practicality implies some type of efficiency, which in turn reinforces practicality.

Another characteristic associated with religion concerns belief systems. Because a belief system revolves around ideas that are expressed through rituals or rules of conduct, it upholds a particular point of view. It does not matter what point of view it takes, as long as it promotes its version of the "truth," that is, the ideas that it preaches. It does not matter if this truth is contradicted by the truths of other religions, such as resurrection in Christianity is contradicted by reincarnation in Hinduism; it only matters that its truth is promoted. It also does not matter if this truth is contradicted by reason, such as the notions of original sin or virgin birth. All of the individual parts of a religion eventually converge into a coherent whole after some period of time, even after some considerable period of time has passed. Because religious beliefs would have some systematic approach to religion, it would have some type of organization.[22] Since all religions develop their own belief systems, a belief system is a characteristic of religion.

As has been said many times before, religion tends to be conservative in its outward form. Of course, all religions evolve, even those that appeared several millennia ago, but that which satisfied ancient believers will not satisfy modern ones. The world has changed, and the changes that have occurred have affected religions in many ways. No matter how long it may take, some old religions will either be replaced by new ones or will be transformed by them. Since the past and present ages are characterized by many religious forms, new religions will likely appear in future ages. It makes no sense to expect that the religions from the axial period will be perpetuated into the future *ad nauseum*, especially when their conceptions of God or gods are conditioned by antiquated historical or cultural factors. Hence, we should conclude that historical facts may be less meaningful to religion than cultural appropriations of the past.

Of the five characteristics associated with religion, three are valid: sociability, practicality, and a belief system. Religion is neither causal nor rational. Sociability forms the social framework of any religion. Practicality assures that religion will retain its worldliness, and a belief system concerns the collected ideas that are preserved in a religion's organization. It should be apparent that these characteristics can lend their support to a religious feeling, but it should also be apparent that a religious feeling is fundamental to these characteristics.

If we were to venture an evaluation of religion in reference to these characteristics, we could say that it supports a social environment in which a religious feeling manifests itself. It promotes individual efforts through group activities, especially the maintenance and perpetuation of social norms within human culture. That it is practical is demonstrated by its efficiency, which is related to the organization of its belief system. Although its effects are evident in its applications, its cause eludes us. If religion is based on religious feeling, then it emanates from within us. It originates within the individual like all other conceptions, but a religious feeling would have to preexist any religious notion, thereby excluding ideas such as the sacred and profane, or the notion of the soul. Religion, therefore, is based on a feeling that originates individually, but has social significance. How a society or age perceives itself will influence the type of religion it develops, or whether it develops any religion at all. We can define the meaning of different religions, but it is difficult to define religion in general. Nevertheless, the characteristics of religion point the way to a definition. At the very least, we can say that religion concerns an idea that is conceptualized internally and is rendered real in the world. Therefore, if we were to render a definition of religion we may say that it is an outward expression of an inner impression that exemplifies a self-reflective social mechanism in which humanity imposes an interpretation on the world. This definition does not confirm the veracity of religious beliefs, but it does emphasize its importance to human culture.

NOTES

1. J.F. Niermeyer, *Mediae Latinitatis Lexicon Minus*, ed. C. van de Kieft, Leiden: E.J. Brill, 1976, s.v. *religio*.

2. The fifteenth century Florentine philosopher Marsilio Ficino (d. 1499) published a book on the Christian religion in Italian in 1474 (*Libro della Cristiana religione*) that was originally written in Latin and later published in that language in 1476 as *De religione Christiana liber ad Laurentium Medicem*. This work is the first known book to refer to religion in the modern sense, and was published as an apology to explain the Christian religion against Judaism and Islam. See *Supplementum ficinianum. Marsilii Ficini florentini philosophi platonici opuscula inedita et dispersa*, ed. Paul Oskar Kristeller, 2 vols., Florence: Leon S. Olschki, 1937, I, pp. lxxvii-lxxix. Also see Michael J.B. Allen, Valery Rees, with Martin Davies, eds., *Marsilio Ficino: His Theology, His Philosophy, His Legacy*, Leiden: Brill, 2002, p. 46.

3. James Martineau, *A Study of Religion: Its Sources and Contents*, 2nd rev. ed., 2 vols., Oxford: Clarendon Press, 1900, I, p. 1. James Bissett Pratt, *The Religious Consciousness: A Psychological Study*, New York: Macmillan, 1920, p. 2. Friedrich Schleiermacher, *The Christian Faith*, 2nd Ger. ed., trans. H.R. Mackintosh and J.S. Stewart, Edinburgh: T.T. Clark, 1928, pp. 12-18. F.H. Bradley, *Appearance and Reality: A Metaphysical Essay*, 2nd ed., Oxford: Clarendon Press, 1897, rpt. 1930, pp. 390-394. John McTaggart Ellis McTaggart, *Some Dogmas of Religion*, 2nd ed., London: Edward Arnold, 1930, p. 3. Similarly, see Robert Leet Patterson, *A Philosophy of Religion*, Durham: Duke University Press, 1970, p. 50, which proposes a modification of McTaggart's definition: "Religion is the belief in a harmony between what is judged to be the highest in human nature and the total environment, together with the emotion to which this belief gives rise and the type of conduct motivated by this belief and this emotion." Henri Bergson, *The Two Sources of Morality and Religion*, trans. R. Ashley Audra and Cloudesley Brereton, with the assistance of W. Horsfall Carter, New York: Henry Holt, 1935, pp. 112 and 130. Clifford Geertz, "Religion as a Cultural System," in *Anthropological Approaches to the Study of Religion*, ed. Michael Banton, London: Tavistock Publications, 1966, p. 4.

4. Wilfred Cantwell Smith, *The Meaning and End of Religion: A New Approach to the Religious Traditions of Mankind,* New York: Macmillan, 1963, pp. 128-129.

5. The difference between God and god is emphasis. Since some people would be offended if the word "god" is used, they may select God instead.

6. This idea is challenged by Rodney Stark and William Sims Bainbridge, *The Future of Religion: Secularization, Revival and Cult Formation*, Berkeley: University of California Press, 1985, p. 8.

7. Alasdair MacIntyre and Paul Ricoeur, *The Religious Significance of Atheism*, New York: Columbia University Press, 1969.

8. Not only is religion an ambiguous word, but its derivation is also uncertain. Although many critics accept the Latin verb *religare* to be the most probable root of religion, mention should also be made of those critics who believe *relegere* (to gather or to collect again) is its root.

9. Emile Durkheim, *The Elementary Forms of the Religious Life*, trans. Joseph Ward Swain, London: George Allen, 1915, pp. 236, 239-240 and 253. It is a conclusion of Durkheimian thought that any religion, regardless where it originates, is the means by which society worships itself. It is also a conclusion of this thought that society has its own reality distinguishable from the reality of individuals. The first of these conclusions, that religion is the means by which society worships itself, is similar to the Hegelian idea that religion is the bearer of human culture.

10. Ludwig Feuerbach, *The Essence of Christianity*, trans. George Eliot [Marian Evans], 2nd Ger. ed., London: Trübner & Co., 1854, rpt. New York: Harper & Row, 1957, pp. 12-14.

11. A good place to begin is Rudolf Otto, *The Idea of the Holy: An Inquiry into the non-rational factor in the idea of the divine and its relation to the rational*, trans. John W. Harvey, 2nd ed., London: Oxford University Press, 1950, pp. 6-7 and 25-30, in which the idea of the "numinous" (derived from the Latin *numen* for "the divine

will" or "divinity") or "the wholly Other" ("das ganz Andere") is explained as an awe-inspiring being that can be rationally understood, but is beyond reason.

12. If a concept is a mental formulation on a broad scale, it must not be confused with the importance of the human being as a unity when distinguished from individual parts thereof because the brain is not more important than the human being as a whole. See M[axwell] R. Bennett and P.M.S. Hacker, *Philosophical Foundations of Neuroscience*, Oxford: Blackwell, 2003, pp. 70-72 and 112. Similarly, Maxwell Bennett, *et al.*, *Neuroscience and Philosophy: Brain, Mind, and Language. With an Introduction and Conclusion by Daniel Robinson*, New York: Columbia University Press, 2007, pp. 21-22 and 131.

13. The idea that religion originates from a particular instinct, or is the result of mental concepts that were formed by human evolution is not new. See Henry Rutgers Marshall, *Instinct and Reason: An Essay concerning the Relation of Instinct to Reason, with some Special Study of the Nature of Religion*, New York: Macmillan, 1898, especially pp. 218-246, who discusses the creation of a "religious instinct." The existence of a so-called religious instinct is denied by Pascal Boyer, *Religion Explained: The human instincts that fashion gods, spirits and ancestors*, London: Heinemann, 2001. Boyer attributes religious concepts to the evolution of the human brain that creates mental structures which help humans behave in certain ways. He argues further that these religious concepts would exist whether or not religion exists. These and similar ideas are also discussed by neurobiology. For the latter, see Jean-Pierre Changeux, *Neuronal Man: The Biology of Mind*, trans. Laurence Garey, New York: Pantheon Books, 1985, pp. 134-140, who discusses the idea of pre-representations as a type of infrastructure that manipulates "mental objects" which preexist interaction with the world. A mental object is described by Changeux as the physical state created in the brain by the activity, electrical and chemical, of neurons (p. 137). Also see his *Physiology of Truth: Neuroscience and Human Knowledge*, trans. M.B. DeBevoise, Cambridge [Mass.]: Belknap Press, 2004, pp. 48-61. Changeux says that pre-representations allow a person to make sense of the world. "They are, in short, the source of imaginative activity" (p. 60). Unfortunately, these ideas in general and the categories of neurobiology in particular have been misconstrued by a discipline now called neurotheology, but which is no more than a corruption of biology. The pursuit of neurotheology as an end in itself, such as that promoted, for example, by James B. Ashbrook, *The Human Mind and the Mind of God: Theological Promise in Brain Research*, Lanham: University Press of America, 1984, is an adulteration of science.

14. David Hume, *An Enquiry Concerning Human Understanding*, sect. III, marginal sect. 18-19 in *Enquiries Concerning Human Understanding and Concerning the Principles of Morals*, ed. L.A. Selby-Bigge, 3rd rev. ed. P.H. Nidditch, Oxford: Clarendon Press, 1975, pp. 23-24.

15. Or as Sartre has said, we wish to be God. Jean-Paul Sartre, *Being and Nothingness: An Essay on Phenomenological Ontology*, trans. and with an introduction by Hazel E. Barnes, New York: Philosophical Library, 1956, p. 566.

16. Karl Jaspers, *The Origin and Goal of History*, trans. Michael Bullock, New Haven: Yale University Press, 1953, pp. 1-6 described "the axial period" as a time of pivotal change in consciousness in many regions of the eastern hemisphere and

approximately at the same time, but Jaspers could not find evidence of direct contact among the different cultures influenced by it. He attributes this change to both religious and philosophical thinkers, and includes Judaism (when associated with the prophets), Hinduism, Buddhism, Confucianism, Zoroastrianism, and Taoism, plus pre-Socratic philosophers and Plato. This pivotal change is also attributed, by extension, to Christianity and Islam. Unfortunately, the axial period has been taken out of historical context from the ancient world, and has been applied erroneously to the modern age.

17. We need to ask the question if religions are still tribal in nature. Are they pretenses based on the assumption of possessing the truth?

18. Stewart Elliott Guthrie, *Faces in the Clouds: A New Theory of Religion*, New York: Oxford University Press, 1993.

19. G. Dawes Hicks, *The Philosophical Bases of Theism*, London: George Allen & Unwin, 1937, p. 34.

20. Michael A. Persinger, *Neuropsychological Bases of God Beliefs*, New York: Praeger, 1987, pp. 16-19.

21. At the same time, religions are necessarily exclusive because they contain innate antagonisms to other religions.

22. It should be noted that the idea of organization itself was questioned by some religious thinkers of the past. Protestant reformers of the sixteenth century, for example, equated the organization of the Catholic Church with false religion because they envisioned true religion to be the worship of God alone without any external apparatus. Because Protestantism emphasized the worship of Jesus Christ more than his message, it emphasized Christology as a special form of Christianity.

Chapter Two

Christianization of Technology

INTRODUCTION

Regardless of the cultural setting and irrespective of historical context, technology influences established religions because it is used by them to convey their meaning. The use of technology is evident in the earliest religions, as it is evident today. It is manifested in statues and vestments, incense and catechisms, stained glass windows and minarets, candles and prayer wheels, bells and organs, altars and scrolls, but it is also made up of religious techniques and ecclesiastical methods. It is both material and procedural. More importantly, however, the opposite reveals a greater truth: of all the religions, it is Christianity that has the strongest influence on technology. This truth is evident everywhere, and it lends its support to the idea that Christianity's influence on technology may be its most significant contribution.

An introduction to Christianity and the latter's impact on technology necessitates an introduction to Judaism that perpetuated the idea of a secular messiah who as a king would create a sovereign state for Jews, commonly known as the kingdom of God. This sovereign state was presumed to be able to resist the influence of foreign invaders; not so much by military might, but by obedience to God's will. The kingdom of God dealt with a preoccupation with present time and an involvement with worldliness that would bring to completion its goal. Since Judaism conceived that its true redeemer was God himself, its idea of a messiah concerned the reign of a king ("the anointed one") who ruled over an earthly kingdom. Although the idea of the Jewish messiah originated as a political leader, it eventually acquired eschatological characteristics.[1]

As commanding as these ideas came to be, they were modified by Christianity. Judaism's concern about a secular leader, however eschatological its function may have been, was transformed into a religious teacher, who was believed to be fulfilled by Jesus Christ and who was considered to be the messiah. Because the crucifixion and death of Jesus Christ did not lead to the immediate end to the world as predicted by its eschatology, a different idea was substituted. This idea concerned the continuation of the mission of Jesus in the ministry of the Holy Spirit, which the early Christian church (later the Catholic Church) was believed to represent, that is, the mission of Jesus Christ was associated with the Catholic Church because the latter signified the head of the Christian community. Early Christianity, therefore, changed the focus of redemption and magnified the importance of eschatology. What resulted was a new religion in which Jesus Christ was perceived to be the fulfillment of Jewish law.

The greatest supporter of the early church was Augustine, whose *City of God* explained why the church was indispensable for the survival of Christianity and why it was necessary for human salvation. The Augustinian view maintained that the kingdom of God continued with the establishment of the early church, whose end could not be predicted. Quoting the Acts of the Apostles (1:6-8), Augustine said that we cannot know the time when the kingdom of God will be fulfilled, or when the end of the world will occur.[2] Although the Catholic Church had an enormous influence throughout the Middle Ages, its authority was seriously challenged in the later Middle Ages by the rise of secular states (notably, England and France), the weakening of the papacy, and the transformation of medieval society. The Protestant Reformation that began in the early sixteenth century intensified its opposition to the papacy. Despite these changes, Christianity created a powerful tradition, whose process of acculturation is known as Christianization.

Christianization was also influenced by Christianity's unique interpretation of history in which Jesus was presumed at some future date to return to the earth in order to judge the living and the dead, after which time history would be completed and the world would come to an end. The period of time from the death of Jesus to the end of the world as described in the last book of the New Testament (known as the Apocalypse of John the Apostle, or the Book of Revelation of John the Divine) was to occur within 1000 years. Although early Christianity was permeated with an apocalyptic spirit[3] in which the world was predicted to come to an end, "the pursuit of the millennium" that inspired so many people remained a myth.[4] The eschatological view that described the end of the world was radicalized at certain times by various medieval and modern thinkers, particularly in times of crisis. The uncertainty of the world necessitated an explanation that made sense for those times when security was lacking. How much this Christian eschatology, or its radicalization influenced Western civilization is difficult to say, but we do know that since the death of Jesus, the idea of the millennium has persisted into the modern age where it has been secularized. It has become a fundamental feature of twentieth century fascism and communism, and has given rise to the idea of inevitability, of a preoccupation with a destiny that cannot be altered.

The influence of Christianity's process of acculturation, or Christianization, as discussed above, has a direct bearing on technology. In point of fact, we should ask how Christianity is able to influence technical methods used to achieve practical results, that is, how does Christianity have an impact on the systematic treatment of all practicalities? How does Christianity affect the procedures, ideas, techniques, devices, products, and underlying mentality of technology? How has Christianity become the greatest force in Western civilization? Of the three fundamental causes of Western civilization, Christianity surpasses the influence of the remaining two: Greco-Roman civilization and Germanic heritage. If the Middle Ages were less religious, the influence of Christianity would have been diminished. Nevertheless, Christianity's influence cannot be ignored. To various degrees, its influence is evident everywhere.

Does the Christianization of technology describe a religious, cultural, or metaphysical process? Although Christianization is a religious idea, it is nonsensical to assume that Christianity has any religious influence on technology. Christianity's influence either originates within a culture which is an individualized example of being that acquires metaphysical significance, or it originates within being which can be analyzed metaphysically that acquires cultural significance. Since culture is a manifestation of being that reflects distinct patterns of social behavior because of particular human conditions, the latter origination described above is the situation that applies to technology. Therefore, the Christianization of technology is a description of a metaphysical process. Because a process is an application, it concerns more than innovative ideas. It concerns the very nature of innovation itself, and innovation means adaptation.[5]

We can understand how technology became Christianized by analyzing those conditions that are associated with Christianity's historical development, particularly, those that have influenced technology during the Middle Ages. Although Christianization can be analyzed from many perspectives, one perspective concerns the interrelationship of those conditions that reveal a truth held in common. And these conditions are essentially three in number: the biblical background to Christianity, the linearity of time, and monasticism. Together, they indicate how medieval society transformed Christianity, and in turn how Christianity transformed technology. This analysis indicates how technology evolved through the Middle Ages, that is, how people exercised their freedom in regard to technology and made choices.[6] As we said above, the Christianization of technology is a description of a metaphysical process.

THE BIBLICAL BACKGROUND

Apart from the description that God, as creator, may be compared to a potter, craftsman, or builder,[7] there are a number of ideas that originated with Judaism that ultimately helped to influence the Christianization of technology. These ideas appear throughout the Old Testament, and by extension throughout the New Testament. They are indicative of the originality of the Jewish interpretation of history. Although there are parallels to some of these ideas with Greek philosophy, it is doubtful whether Judaism had any influence on the Greeks. In fact, the opposite is more likely to have occurred, especially when Christianity became Hellenized during its formative years. The biblical background to the Christianization of technology is based on four main ideas.[8]

The first of these ideas concerns the creation of the universe. The first sentence of the book of Genesis, that is, the first book of the Old Testament, describes that God created the universe, enumerated as the heavens and the earth. Therefore, it posits creation to a willful and divine agency that also became the God of the Jews. God is distinguished from other contemporary deities that are presumed to be inferior. Because the Jews do not consider the earth to be eternal, it must have a distinct beginning (Genesis 1:1). Judaism presumed that creation is an indispensable event because it has been assigned the greatest value.

The second idea describes how humans are made in "the image of God," and it is an extension of the first idea (Genesis 1:26-27 and 9:6; Wisdom 2:23).[9] The creative powers of God are now associated with humans, but humans have these powers only because they have received them from God. Being made in God's image is offered as an explanation where human creativity originates (Sirach 17:1-3; 1 Corinthians 11:7 and 15:49; Colossians 1:15 and 3:10). It also presumes that these creative powers will be applied. Explanations of this creativity are followed by further elucidations that describe how these powers are to be used, powers that concern the utilization of the earth and all forms of life on it (Genesis 1:26-30 and 9:7). Therefore, these creative powers presuppose the application of many techniques, and it assumes that they are features of a practical matrix. These powers are existent despite the idea of human imperfection associated with so-called original sin that resulted from Adam's act of disobedience.

The use of the earth implies materialism. One might think that materialism opposes religion by placing too much emphasis on physical comforts when compared to spiritual values, but materialism may also be the effect of simply using the world for well-being. All religions originate in the world, and some religions, such as Hinduism and Buddhism, make use of the world in order to achieve perfection, but they do not involve worldliness. Judaism and

Christianity, on the other hand, act differently because both of them make a commitment to the world and both require the use of the world for human betterment. They are actively involved with the world, and therefore anticipate the world's technologization more than other religions.

These creative powers are related to a third idea. In addition to the use of the earth, humans are also expected to help God in his divine plan of creating a community that will lead to a heavenly kingdom (Sirach 38:34). This idea is based on the notion of the kingdom of God that describes the goal of the Old Testament Jews in establishing a sovereign state independent of foreign domination. The kingdom of God represents the triumph of the belief in God and the resultant reward he makes to his believers. Christianity transformed this notion. It interpreted this notion religiously, not politically. The Christian interpretation of the kingdom of God (*regnum Dei*) redefines the Jewish idea of the heavenly kingdom (*regnum caelorum*), or the kingdom of the Lord (*regnum Domini*) that will begin in this world, but will be fulfilled in the next.[10]

The fourth idea concerns the end of life. It says that life on earth will come to an end that will bring to completion God's earthly plan (John 9:4; Ephesians 5:16). The end of life signifies not only human life, but all life. Therefore, the earth will be destroyed, and in the next world, the kingdom of God will be fulfilled. Christianity, as mentioned above, transformed the Jewish interpretation of this kingdom because Christianity makes a projection into the future. Judaism emphasized the present over the future, but did not advocate leaving the world. The Christian interpretation emphasized that the world (perceived to be temporal) has a distinct beginning and a definite end, both of which are subject to a linear (and final) interpretation of time.

All of these ideas are religious in origin. They may touch on politics or economics, but they are essentially concerned with the role of religion in society. They deal with the place of God in history, why his law must be obeyed, and what influence his law has on human culture and morality. These ideas are also illustrative of the importance religion has played among the Jews and how they have given the concept of God and the creation of the world the highest value.

When transposed to technology and when accepted as essential features in a technological society, these biblical ideas signify the following: 1) humanity's intended means, as encouraged by God, will lead to the use and exploitation of nature; 2) the intended end, which is the practical effect of the idea of the heavenly kingdom, is humanity's improvement through technology in product, procedure, and design; and 3) both the means and the end will occur through the linearity of time that anticipate the perpetuation and fulfillment of technology's purpose.

Although Christianity was proposed as a religion that was not involved with the world, eventually it became a religion of the world. This involvement led to Christianity's secularization, particularly when the Catholic Church fell victim to the demands of politics. The church as the spiritual leader of Christianity became associated with many struggles with worldly matters that had little to do with salvation. Because Christianity used its religious calling for secular concerns, it crossed the threshold into worldliness. Nevertheless, the idea of the foreordained events described in the Apocalypse continued to have an influence.

These ideas are related to other ideas that supplement them, and one such idea concerns the role of Christianity in the idea of progress. Although there were early Christian writers who perpetuated this idea,[11] the belief that Christianity is a progressive religion remains a myth. The thinkers who associated Christianity with the idea of progress were concerned with more than the salvation of souls. Apart from the opposition to progress by Augustine, circumstances persisted which allowed Christian writers to lend their support to a progressivism that presupposed advancement in society and morality. It was tied to the biblical idea of helping God in his divine plan of creating a community that will lead to a heavenly kingdom, and was also transposed to life in general.

The promoters of progress fostered a belief that Christianity was more than just the ultimate religion that contained the only means to salvation. They also fostered a belief that the world would get better because believing in Jesus was thought to be the means that fostered a positive view of life. The gospels were known as the "good news" because they describe how happy life can be if Jesus is accepted as "the Son of God." Without such a view, the persecutions of the early Christians would have been meaningless. As a result, these persecutions were interpreted as verifications of Christian belief and proof of the truth of Jesus Christ. Christian progressivism gave them a cause for living, and a reason for dying. This view is not far removed from the influence of progressivism on technology. Without it, Western civilization as it was influenced by Christianity would be less formative, dynamic, and innovative than it came to be.

We should also question whether the Christianization of technology itself is based on a myth. If Christianity is concerned only with salvation, then everything else, including progress, is superfluous. But the religious and personal requirements for salvation gave way to the idea of human improvement, that is, the notion of progress resulted from those steps that have been used for salvation. Therefore, the idea of progress, apart from other factors, was influenced by the belief in Christianity, and this belief persisted apart from the fact that progress was directly opposed to Christianity.[12] The Bible has instructed humans to use nature as the means of creating a hospitable environment for themselves that was interpreted originally as the means to creating the heavenly kingdom, but later as the means to progress. We need to reflect whether or not the Christianization of technology is a myth, but we should also accept the reality of technological progress. Although the cause might be imaginary, the effect is real. The Christianization of technology deals with the power of ideas, and technological progress deals with the effect of their application.

LINEARITY OF TIME

Although it can be said that the Old Testament emphasized the notion of present time because the Jews were concerned about the preservation of tradition, it may also be said that Jewish tradition came to become more futuristic and therefore acquired eschatological characteristics. Furthermore, Christianity that originated as a radical Jewish sect could not have been established as a religion without substituting the future for the present. It conceived that time was laid out rectilinearly from an absolute beginning to an irrevocable end, and implied that the power of God was omnipotent, whose power would be applied to all existence, even though linearity in itself does not prove eternity. Since linearity differentiated Christianity from other religions, linear time penetrated every aspect of Christian culture.[13] It pointed ultimately to future events that have not occurred, and predicted that Jesus would return to the earth in order to judge the living and the dead (the Second Coming). This return would occur only once, as the crucifixion and resurrection of Jesus were also singular events.

Of time's three variants (the past, the present, and the future), the future became the principal variant of Christianity, and it is especially evident in the Apocalypse of John the Apostle—the last book of the New Testament—that described events believed to be foreordained. Hence, Christianity became an apocalyptic religion. Since the future was anticipated, it was thought to be the completion of the process that would lead to the ultimate reality: the end of the world and coexistence with God.

Linearity permanently replaced the cyclical conception of time characteristic of the pre-Christian world in which everything was thought to be repeated without beginning or end with a new conception that began with the first coming of Jesus to be followed by the projected Second Coming and the end of the world.[14] The whole purpose of Christianity concerns the salvation of souls, a goal that would be pointless if time was not laid our linearly. It should be evident that linearity perpetuates Christianity's success because it promotes optimism, fosters progressivism, and encourages the pursuit of perfection. Although cyclicity reveals a recurring succession of events, linearity reveals the inherent augmentation of them laid out to their final end. The succession of increments of one's life, or culture, or civilization that progresses toward a continuum characterizes the linearity of time. Of course, cyclicity in the pre-Christian world does not mean that there were no people who could conceive of time linearly, nor does it mean that they lacked the idea of progression. Prior to Christianity, linearity had little meaning in a culture where cyclicity ruled the day.

These ideas have influenced technology because the future is also technology's principal variant. Although we live in the present, technology is projected toward the future,[15] but technology's future is not Christianity's future. It is for this reason that we can say that Christianity's influence on technology is not religious, but metaphysical. Since the essence of the future is possibility, the latter becomes a characteristic of technology, which is open-ended, uncertain, and undefined, but Christianity's future is closed-ended, certain, and defined.[16]

Any conception of time supposes some type of time awareness. Although the conception of time is an abstraction, it is still a feature of technology because it deals with the practical effects of time reckoning. Hence, the reckoning of time, not just its awareness, is a feature of technology, either one that repeats itself (cyclical), or one that does not (linear). When any type of time reckoning is emphasized, time becomes technical, and linearity is a more sophisticated representation of this effect. Linearity is also a representation of a specific conception of time that concerns the imposition of an idea not supported by nature. It is an example of humanity's demands on the natural rhythm of time.

Nevertheless, the future and technology have characteristics in common because of Christianity. The future, which lacks both reality and actuality, expresses more than a concern about possibility. It is also occupied, like Christianity, with the ultimate effect of time's linearity, that is, technology has been positively impacted by Christianity's forwardly directed ideas. It posits the future that we have set before ourselves.[17] In a sense, we can say that Christianity facilitates technological change because technology concerns the application of its means to a projected end. Although technology is a continuum of self-perpetuating means that introduces change as an effect of its application, technology is still occupied with the idea of a projected end implied in its means. All cultures possess technology, but Christianity has the peculiar ability to augment this relationship. It has helped to intensify the future as the chief variant of technology.

MONASTICISM

Western inventiveness and its innovative applications are the result of Western civilization, but the West has also assimilated inventions from other civilizations, many of them from Asia. These include the stirrup, horseshoes, the wheelbarrow, and paper. In addition to inventions, the West has assimilated many ideas, techniques, and beliefs from non-Western peoples by means of cultural diffusion. One of these concerns Christian monasticism, which originated in late third century Egypt, and was adopted in the West in the following century.

Regardless of where it appears, monasticism presupposes a discipline rarely found outside of a cloistered, and allegedly isolated environment. Because the laity lacked this discipline, it looked to the monasteries for direction and learned how to live a Christian life by observing the lives of monks. In a sense, we might say that the only type of truly religious life that was available to Christianity in the early Middle Ages was monastic. Since monasteries acquired an importance far in excess of their numbers, they had a positive and lasting influence on the development of Christianity as it evolved through the Middle Ages. A life devoted to prayer and a willingness to follow a highly regulated lifestyle were important monastic characteristics.

Although Western monasticism shared many characteristics with those types that originated in the Eastern provinces of the Roman Empire, the latter maintained a more ascetic and more contemplative approach than Western monasticism. By contrast, Western monasticism, apart from its shortcomings, was noted for its more comprehensive rules of orderly conduct, its pursuit of self-control, its highly efficient regimentation of a monk's daily routine, and the ideals of a community that placed supreme emphasis on obedience. Beginning in the fourth century and improved in the sixth century by the rule by St. Benedict of Nursia, Western monasticism had also assumed a more inwardly directed motivation, a greater level of meditation, and an improved self-discipline that was later transformed by the religious changes of the eleventh century.[18] Monasticism promoted a life shared in common with other monks who followed a rigid routine based on strict discipline that described the practical steps considered to be essential for the achievement of its goals.[19]

Monasticism in general and the Benedictine order in particular, both of which devised very inwardly directed motivations, required additional regulations that affected every aspect of a monk's life. The participation of monks in the celebrations of saints' days, of which there were many, mandated the creation of customaries or monastic constitutions that described all the activities in great detail, both liturgical and organizational, of a monastic community. Despite the influence of regional or papal politics, Western monasticism provided an environment in which technology and its applications could flourish, artistic creativity could be promoted, and learning (both within and outside monasteries) could be encouraged, and all of these activities helped to reinforce a monastery's self-sufficient directiveness.

Either because historical conditions in the early Middle Ages favored monasteries as isolated communities, or because these conditions created circumstances that encouraged technology, monasteries were responsive to the application of what is known as the useful arts that facilitated technological development. Apart from the invention of the mechanical clock that helped monks become more punctual for religious services and is attributed to the late thirteenth or early fourteenth century, monasteries were known for emphasizing the importance of manual labor that included all types of practical work, in addition to agricultural cultivation.[20] Manual labor also included the application of those useful arts that helped monastic communities become independent. The useful arts may even have been fundamental to the development of manuscript illumination that is indicative, in part, of the technical expertise needed to produce and perpetuate some types of art at that time. Despite the ambiguities associated with manual labor, its association with the useful or practical arts should be assumed.[21] More importantly, however, the pursuit of the useful arts requires the organization of rules and procedures, without which a craft or skill cannot evolve into a discipline. This organization became one of the greatest attributes of monasticism. Its methodology of work, study, prayer, and purposeful dedication reinforced the practical effects of Christianization.[22]

The Christianization of technology was also influenced by the characteristics associated with the method utilized in medieval monasteries. If technology is not comprised of products and devices of technical organization, then it must be comprised of methods, and a method is a multilayered structure that a procedure needs in order to persist in the world. Apart from the fact that a method is an essential characteristic of technology, it is also made up of individual techniques that concern many different procedures. The fact that technology is made up of methods has a bearing on monasticism because the latter developed a method of salvation that subjugated worldly matters to otherworldly goals. Whether its method was successful in saving souls is not known, but it certainly was successful in making the monastic lifestyle an organized and efficient reality. Individual monks most likely internalized this method, and once internalized, it was applicable to many pursuits. The method of monastic discipline comprised several techniques that included a daily schedule, tied to the hour of the day, for work, study, prayer, and other activities that were considered to be useful for the perfection of individuals and the achievement of salvation. Because the Second Coming was delayed, perfection and advancement became the new goals of Christianity. Therefore, the greatest attribute of monasticism did not concern the preservation of learning, however important it became in the Middle Ages, but the perpetuation of a method that potentially encouraged all types of organization.[23] Initially proposed as a way of life for monks, monasticism gave medieval society, both within and outside monasteries, an organization that emphasized systematic procedures. It described a process that was arranged into an ordered whole and concerned structure in which individual features were tied to a common goal. Monasticism gave Christianity a method that set in place its practical effects. It helped to combine theoretical thought with empirical investigation.[24]

This method is also associated with the idea of a work ethic. More than the advocacy of the usefulness of manual labor, the work ethic became a feature of the monastic method. Because an ethic means a discipline by which one acts, it becomes a part of a system of values. In moral philosophy, ethics means the principles of moral conduct. It is by extension of these ideas that we arrive at the description of a "work ethic," which concerns the belief that work is morally good. It should not be confused with the need to work hard, which is a means to survival. Because monasticism utilized a work ethic as part of its rule, it became the basis of the expression "to work is to pray" (*laborare est orare*), that is, work was worship of God when this method was applied.[25] Eventually, it was also associated with the development of technical knowledge, that is, with the skills needed for the pursuit of the useful or the mechanical arts.

CONCLUSION

The three conditions discussed in this chapter are important not only because of their individual relationship with the development of Western technology, but also because of their interrelatedness and simultaneity. Any one of these conditions is important in itself, but collectively they reveal an influence that has set Western civilization on a course that is dynamic, compelling, and decisive.

The first condition concerns the biblical background of Christianity. This condition is meaningful because humanity is presumed to be made in "the image of God," which implies that humans share some characteristics with God, particularly, those creative powers that are to be used in God's plan of establishing a community that will lead to the heavenly kingdom. When applied to technology, "the image of God" augments the idea that technology is dependent on creative activity generated by an inventive spirit. The biblical background offers an explanation where this inventive spirit originates. Because it defines creativity as an attribute of God, its application implies God's approval. A denial of the significance of this "image" questions God's will. Therefore, technology is a beneficiary of the idea of "the image of God."

The second condition concerns how these creative powers received support from a linear conception of time that is perceived to be laid out in a straight line from an absolute beginning to a final end. It is because of linearity that Christianity became an apocalyptic religion. It posited the future as the fulfillment of the present and made Christianity the vehicle of a process that was thought to fulfill the prophecies of the Old and New Testaments. This process promoted an optimism that advocated progressivism and encouraged the ideals of perfection. Even when faced with adversity, Christianity gave a positive attitude that fostered hope.

These ideas have influenced technology, in part, because the future has become technology's principal variant of time. Although linearity is an essentially unnatural interpretation, technology is concerned, much like Christianity, with the application of its means to a projected end. Linearity is the result of a process of self-perpetuating means because its projected end is inherent in its means. Therefore, Christianity has bestowed on technology a forwardly projected directiveness.

The third condition relates to monasticism, not so much because of its original intent as a way of life for the attainment of salvation, but because of its method that introduced an organizational structure into Western society. Originally important for its rules of orderly conduct that encouraged self-control, it evolved into a highly developed motivation that promoted perfection. The greatest effect of its method concerned the practical steps needed for the attainment of its goals. Eventually, monastic rules and procedures were organized into a system that encouraged the development of various crafts, that is, the monastic methodology of work, study, and prayer were applied to many useful pursuits. Christian monasticism perfected a method that encouraged the combination of speculative thought with practical applications. This method evolved into a practical matrix that influenced the development of technology.[26]

Not only has the biblical background generated the possibility of other conditions, these conditions are also important because they are interrelated. In the early Middle Ages, this interrelatedness also entailed simultaneity. Together they illustrate why Christianity's metaphysical conditions are more important than its historical. Cultures manifest the exercise of human freedom as an ontological characteristic, and when choices are made, that is, when humans reveal this freedom, we may analyze the metaphysics of culture. The Christianization of technology as expressed through these conditions reveal a metaphysical process that has strengthened and directed the development of technology in the West. It has enabled technology to acquire essential characteristics that may help it evolve into a religion.

NOTES

1. S[igmund] Mowinckel, *He That Cometh*, trans. G.W. Anderson, New York: Abingdon Press, n.d. [1954], pp. 4-9.

2. Augustine, *The City of God against the Pagans*, ed. and trans. R.W. Dyson, Cambridge: Cambridge University Press, 1998, pp. 903-904 (bk. XVIII, ch. 53).

3. An apocalypse may be defined as a divine revelation of the future. It describes humanity's struggle with good and evil, and it is based on the assumption that good will triumph and evil will be defeated. It manifests the completion of a process and presupposes a linear conception of time.

4. Norman Cohn, *The Pursuit of the Millennium: Revolutionary Millenarians and Mystical Anarchists of the Middle Ages*, 3rd rev. ed., New York: Oxford University Press, 1970, pp. 85 and 109.

5. This metaphysical process reflects a cultural situation and the changes that result from it despite the controversy created by associating Christianity (and other traditions) with the domination and exploitation of nature. See William Leiss, *The Domination of Nature*, New York: Braziller, 1972, pp. 29-35. It is also reflective of the idea of humanity's stewardship of nature, as noted in John Passmore, *Man's Responsibility for Nature: Ecological Problems and Western Traditions*, New York: Scribner's, 1974, pp. 28-35. If nurturing of nature is one part of this description and exploitation is another, then usefulness is the result. But a different question concerns who benefits from this usefulness. Do humans benefit when species are protected from extinction? Do humans benefit when they use natural resources to excess? Regardless of the answers to these questions, humans would profit in some way, otherwise, these questions and the conditions they describe would have little meaning. Of course, the destruction of nature would ultimately entail human destruction.

6. Apart from the interpretations of White and Mumford on the importance of Christianity for the development of technology, one idea emphasized by both these thinkers, regardless of their historical or philosophical persuasions, concerns the uniqueness of Christianity as a motivating factor. Mumford lays emphasis on the monastic ideal, and White examines the historical roots to Western technology while also reflecting on its social effects, especially the rational use of nature, the positive attitude to manual labor, and the importance of praxis over theory. These interpretations are based on the presence of Christianity, however varied they may be, but they do not explain how technology acquired the effects of a progressive mentality that intensified the preoccupation with means over ends, emphasizing the predominance of the future over the present and bestowing on technology the distinction of inevitability. See Lynn White, Jr., "What Accelerated Technological Progress in the Western Middle Ages?" in *Scientific Change: Historical studies in the intellectual, social and technical conditions for scientific discovery and technical invention, from antiquity to the present*, ed. A.C. Crombie, New York: Basic Books, 1963, pp. 272-291, and Lewis Mumford, *The Myth of the Machine*, 2 vols., New York: Harcourt, Brace, Jovanovich, 1966-1970, I (1966), pp. 263-267. A criticism of these authors should also include a criticism of George Ovitt, Jr., whose book, *The Restoration of Perfection: Labor and Technology in Medieval Culture*, New Brunswick: Rutgers University

Press, 1987, presents unconvincing arguments of White and Mumford. The historical evidence he cites confirms the likelihood of the opposite of his own contentions, especially in regard to "Man as custodian" (pp. 70-72), "The medieval ethic of co-operation" (pp. 85-87), "The products of twelfth-century labor and technology" (pp. 138-143), and 'The loss of perfection" (pp. 200-202).

7. For example, potter (*figulus*): Sirach 33:13; Isaiah 29:16, 45:9 and 64:8; Jeremiah 18:2-6; Romans 9:21. Craftsman (*artifex*): Wisdom 13:1; Hebrews 11:10. Builder (*conditor*): Hebrews 3:4 and 11:10. Since God was perceived to be a technician, the idea arose that he is a clockmaker because the forces of the universe move with such clock-like precision. The idea of the clockmaker is derived from the fourteenth century scientist, Nicole Oresme (d. 1382). See Edward Grant, ed. and trans., *Nicole Oresme and the Kinematics of Circular Motion: Tractatus de commensurabilitate vel incommensurabilitate motuum celi*, Madison: University of Wisconsin Press, 1971, pp. 294-295.

8. As an introduction to these ideas, it is imperative to consult Ernst Benz, *Evolution and Christian Hope: Man's Concept of the Future from the Early Fathers to Teilhard de Chardin*, trans. Heinz G. Frank, Garden City: Doubleday, 1966, pp. 123-127, although I disagree with some of Benz's selections.

9. Similarly, likeness (*similitudo*): Genesis 5:1; James 3:9, and image (*imago*): Matthew 22:20; Mark 12:16; Luke 20:24; 2 Corinthians 4:4.

10. The New Testament has many references to the kingdom of God, or the heavenly kingdom, or simply, the kingdom.

11. Tertullian, *Apology*, trans. T.R. Glover, Loeb Classical Library, Cambridge [Mass.]: Harvard University Press, 1931, pp. 186-187 (ch. 40, sect. 13). Eusebius of Caesarea, *Ecclesiastical History*, trans. Kirsopp Lake and J.E.L. Oulton, 2 vols., Loeb Classical Library, Cambridge [Mass.]: Harvard University Press, 1926-1932, I (1926), pp. 390-391 (bk. IV, ch. 26). Orosius, *The Seven Books of History against the Pagans*, trans. Roy J. Deferrari, Washington: Catholic University of America Press, 1964, pp. 283-285 (bk. VII, ch. 1). Lactantius, *The Divine Institutes, Books I-VII*, trans. Sister Mary Francis McDonald, Washington: Catholic University of America Press, 1964, pp. 97-101, 376-382 and 393-395 (bk. II, ch. 2, bk. V, ch. 19, and bk. VI, ch. 2).

12. Rudolf Bultmann, *History and Eschatology*, Edinburgh: Edinburgh University Press, 1957, p. 70.

13. Christian culture is not alone in developing ideas of linearity. China also developed two reckonings of time: 1) a cyclical conception that was limited traditionally to sixty years after which a longer period of time was thought to be a violation of nature, and 2) a linear conception that revealed one's place in an endless line of time that connected an individual to his ancestors and his progeny. See Michael Loewe, "Cyclical and Linear Concepts of Time in China," in *The Story of Time*, ed. Kristen Lippincott, *et al.*, London: Merrell Holberton, 1999, pp. 76-79.

14. For cyclicity and linearity, see G.J. Whitrow, *Time in History: The Evolution of Our General Awareness of Time and Temporal Perspective*, Oxford: Oxford University Press, 1988, pp. 47-49 and 51-53 respectively.

15. See my "The Conception of Time and Its Relationship to Technology," *Research in Philosophy and Technology*, 19 (2000), 221.

16. These Christian characteristics of closed-endedness, certainty, and definitude are relevant to time regardless of time's abstraction. It is misconstrued to conclude that time in the New Testament "is never spoken of in an abstract manner," as interpreted by Oscar Cullmann, *Christ and Time: The Primitive Christian Conception of Time and History*, trans. Floyd V. Filson, rev. ed., Philadelphia: Westminster Press, 1964, p. 54.

17. Erich Frank, *Philosophical Understanding and Religious Truth*, New York: Oxford University Press, 1945, rpt. 1966, p. 68.

18. In general, see [Edward] Cuthbert Butler, *Benedictine Monachism: Studies in Benedictine Life and Rule*, 2nd ed., London: Longmans, 1924, rpt. Cambridge: Speculum Historiale, 1961, and David Knowles, *Christian Monasticism*, New York: McGraw-Hill, 1969.

19. Apart from the Cistercians who excelled in agriculture and industry, monasticism began to decline in the eleventh century, aggravated in part by the weakening of the monastic ideal and by the appearance of mendicant friars, notably, Franciscans and Dominicans, who were traveling preachers serving the sick, the ignorant, and the destitute. Since the friars had no fixed place to live, they begged for food, hence, the name "mendicant." Although monasticism was perceived to be the best means to salvation, this idea was no longer reserved exclusively for monasteries.

20. Except in agriculture and business, manual labor was discouraged in Greco-Roman antiquity and was performed almost exclusively by slaves. This negative view did not prevent its inclusion when monasteries appeared, but eventually, agriculture creased to be a normal activity for medieval monks, and was performed primarily in monasteries by free or slave laborers. Manual labor of non-agricultural work characterized the monastic work ethic and became the basis of the useful or mechanical arts. Manual labor in the Benedictine rule is discussed in chapter 48 (*De opera manuum cotidiana*). See Adalbert de Vogüé and Jean Neufvillle, eds., *La règle de saint Benoît*, 7 vols., Paris: Éditions du Cerf, 1971-1977, II (1972), pp. 598-605.

21. Dom John Chapman, *Saint Benedict and the Sixth Century*, New York: Longmans, 1929, pp. 171-172 assumed that the fine arts, such as cabinet making, stone carving, or metal working, were included in the various crafts utilized in monasteries. [Edward] Cuthbert Butler, "Saint Benedict and the Sixth Century," *Downside Review*, 48 (1930), 194 discusses Chapman's assumptions, and concludes that only the useful, not the fine arts, were meant.

22. Herbert B. Workman, *The Evolution of the Monastic Ideal from the Earliest Times down to the Coming of the Friars: A Second Chapter in the History of Christian Renunciation*, London: Epworth Press, 1913, rpt. Boston: Beacon Press, 1962, pp. 156-161.

23. Jean Gimpel, *The Medieval Machine: The Industrial Revolution of the Middle Ages*, 2nd ed., Aldershot: Wildwood House Ltd., 1988, p. 47.

24. A.C. Crombie, *Augustine to Galileo*, 2nd ed., 2 vols., London: Mercury Books, 1961, I, pp. 175-176.

25. The importance of manual labor and its place within the development of a work ethic is evident in Theophilus' *De diversis artibus*, a treatise on the useful arts written during the first half of the twelfth century. Theophilus was a Benedictine monk and a metalworker who described the techniques associated with many crafts that included painting, glazing, and metallurgy. *De diversis artibus* is an example of the Benedictine pursuit of the useful arts, and it is the culmination of this pursuit. See Theophilus, *The Various Arts*, trans. with Introduction and Notes by C.R. Dodwell, London: Thomas Nelson & Sons, 1961. Particularly noteworthy is the preface to the first book of this treatise that encourages a heavenly reward for those who avoid idleness, pursue some useful occupation, and contemplate new things (*ibid*., p. 1).

26. Like other religions, Christianity contains contradictions. One of these concerns the conflict that results between its endorsement of individuality because salvation is pursued individually, and its presumption that it presents a belief, that is, a framework, reinforced by institutions to which individuals are expected to conform. This conflict is thought to be resolved when individuals accept their role within Christian society. Monasticism is an example in which the pursuit of individuality was replaced by communal activity and obedience. It subordinated individuality to obedience that was essential for the perpetuation of Christian culture, and by extension, for the perpetuation of technology. The agent of this perpetuation was the monastic method.

Chapter Three

Secularization and Its Effects on Technology

INTRODUCTION

If a religion is an outward expression of an inner impression that exemplifies a self-reflective social mechanism in which humanity imposes an interpretation on the world, then religion is extremely important to human culture. In addition to promoting a belief system, a religion also promotes socialization, a process by which individuals are acculturated to society. As soon as we encounter socialization, we have crossed the threshold into values and valuation. And since new generations of believers are introduced to the beliefs and rituals of a religion, they are regarded to be capable of perpetuating that religion. Therefore, a tradition is reinforced that is passed down from one generation to another, but the presence of religion and the traditions it creates may be weakened by the process of secularization, which itself has contradictory results. Although the weakening of religion has an effect, this effect may not be negative.

Since most cultures promote some type of religion, the weakening of religion may be uncommon. Whether genuinely derived from human culture or artificially imposed on it, the importance of religion is proportionate to the relationship it has to its respective culture. But the promotion of religion is in inverse proportion to a culture's prosperity, that is, poor cultures or countries are generally more religious than rich ones. Since secularization occurs simultaneously with the development of modernization, which to some extent may be equated with westernization, it signifies that religion may be undermined by changes that are taking place within society to which religion is connected. If a culture disregards or belittles the influence of religion, then that culture may make something else into a religion, but there may be residual emotions that persist after religion has been demoted. As we said above, the absence or denial of religious beliefs may prepare the way for new beliefs, and it is this condition that the present age must deal with. But future religions are to be distinguished from civil religions that may be defined as types of popular (or folk) religions of the state that unify its citizenry to prescribed values, such as patriotism.

As already mentioned, secularization implies a weakening of religiosity. Although it may be assumed that religion would be unwelcome, the opposite may be the case, not that old religions would be reinvigorated, but that new religions may come into being. The weakening of religion is only one part of the process of secularization. Another part concerns opening up new vistas of activity. The weakening of religion may support the view that secularization is a theory—however loosely defined—not an ideology. Nor is it a concept, but a signature of what has been.[1] Secularization is a reflection posed backwards, indicating a state of being that is not at hand.

It should be apparent that the world is undergoing some type of secularization, although the final result cannot be known. If secularization were an ideology, it would already follow certain doctrines or beliefs that constitute a system, but this condition is not evident because the circumstances that may give rise to new religions are themselves still forming. Secularization is the result of the socialization of certain factors that manifest the effects of social differentiation, and these factors can be explained historically.[2]

The persistence of religious beliefs seems to confirm some type of stability, but this persistence does not prove that existing religions will survive, since what people believe and how they express their beliefs are two different things. Secularization is a way of perceiving the world, but it may not be the way the world actually is. Even the privatization of religion may exist while society becomes secularized. Of course, religion in the modern age may become entirely personal, but this is not to say that it cannot be used for public purposes, such as the compelling influence imposed on believers to vote and act conservatively, or in some cases, to act negatively or self-destructively, the latter of which is an extreme reaction to secularization. Therefore, the privatization of religion may still have an effect. So the issue is not over the privatization of religion. It is over its expression in society, in whatever form it takes, although privatization cannot account for new religions that may appear. Regardless of the status of existing religions, the transformation of religious beliefs will not be immediate. It may take some time for existing religions to be replaced by the new, but this replacement can take place.

In general, we may say that secularization describes a process that is applicable to society in general, but it has a lessened influence on religious organizations that are already established and that persist in their beliefs. Therefore, the Muslim clergy, or the Catholic Church, or Buddhist monasticism may continue with their traditions while their respective societies are secularized. Because secularization has an influence on society, it falls prey to the application of government within society. Apart from the suppression of religion by some governments, conservative governments may use religion for political ends, and liberal governments may attempt to replace religion with politics. The conservatism of a government reinforces religious traditions, but liberalism may be used in contradictory ways, one being its opposition to government, yet another in its support of it. Although not originally intended to displace religion, secularization may achieve this by substituting something else for it. This substitution often confuses genuine religions with civil religions, which are more political than religious. The fact that the influence of religion may increase in a secular age does not prove that secularization is not occurring, or that it is a false theory,[3] but that the final struggles of some religions, and the transformations of others are taking place. It is within this context that technology as a cultural rather than a technical phenomenon may qualify as a new religion.

HISTORICAL BACKGROUND

Secularization may be studied historically. Although the Enlightenment is accredited with being the age in which secularization first appeared in Western Europe, this conclusion is unfounded. In fact, as any medievalist knows, secularization made its first appearance five hundred years earlier in the thirteenth century. The reality of this and later centuries contradicted the moral teaching of the Catholic Church—the predominant religious institution in Western Europe at that time—not only for its views on usury in which it was no longer considered to be immoral to make a profit on the lending of money, but also because political power greatly intensified that put pressure on the Catholic Church to become embroiled in many political disputes. As an institution, the church was forced to become more worldly, particularly in its political struggles with England, France, and the Holy Roman Empire, that is, it was partially responsible for the development of secularization that later became evident. Although the Catholic Church wielded ecclesiastical punishments for political ends, ultimately its actions turned a critical eye onto itself. There were those at that time, such as Marsilius of Padua, who considered the church wrong for assuming and exercising any type of political power, and who upheld the view that the true position of the church concerned spiritual matters only.[4]

Augmented by the Avignon papacy (1309-1378) in which the popes resided in France and were influenced by the secular affairs of the French monarchy, and the Great Schism (1378-1417) in which two popes claimed to be legitimate simultaneously, the power and respect of the Catholic Church further declined. The economic and social problems created by the bubonic plague (the Black Death) that began in the middle of the fourteenth century undermined the social fabric of lay society and the support it received from the Catholic clergy. Although economic conditions improved through the fifteenth century, enough damage had been done to economics to threaten medieval society. The late Middle Ages was also characterized by the gradual replacement of land as the principal source of wealth in an agrarian society—which was the principal requirement of feudalism—with money as the principal source of wealth in a commercial society. An economy tied to landed estates in rural areas gave rise to an economy based in cities that also laid the basis for capitalism. In addition to papal politics described above, this shift in economics also helped to accelerate the development of secularization.

Although the late Middle Ages was characterized by the disintegration of the unity of Christendom, which lent its support to the development of secularization, the latter was also influenced by science, particularly by the Scientific Revolution that first appeared in the sixteenth century and was derived from a heliocentric or sun-centered interpretation of the universe. In the Middle Ages, astronomy was based in large part on the views of Aristotle (fourth century BCE) and later expanded by Ptolemy (second century CE) in which the earth was perceived to be the focus of everything. This interpretation, later known as the Ptolemaic system, upheld the view that the earth was fixed, immovable, and located at the center of the universe, and all other heavenly bodies, comprising the moon, the sun, the planets, and all the stars were believed to revolve around it. Beyond the stars, there was empty space.

The Ptolemaic system was challenged by the astronomical observations of Nicolaus Copernicus (d. 1543), the Polish astronomer who proposed that the earth and the planets revolved around the sun, not the reverse. The observations of Copernicus and subsequent scientists, who stressed astronomy more than cosmology, began a new system of inquiry that characterized the Scientific Revolution. It questioned the science that had evolved through the Middle Ages and the theology of the Catholic Church that was supported by it. It challenged the church's interpretation of science, which was based, in part, on the Bible, an interpretation that predated the development of the scientific method. Although the Catholic Church was often interested in the development of new scientific theories, this interest was meaningful only when church doctrine was not challenged. When the church was forced to make a choice between theology and science, it chose theology. If it could be demonstrated that the church's interpretation of nature was contrary to reality, which meant that the Bible and the ancient authorities that supported similar conclusions were not scientifically reliable sources, then it set in motion a spiral of doubt that questioned other aspects of the Bible, and by extension, the Catholic Church itself. Once doubt began, it was unknown where it would lead, or if it could be stopped.[5] Would other aspects of Christian doctrine also be called into question? Is secularization peculiar to the West, or is it applicable to non-Western cultures as well?

The Scientific Revolution reflected a transformation of Christian society. It was not the culmination of a former tradition, considering that it occurred shortly after the end of the Middle Ages, but the beginning of a new tradition that extended into the eighteenth century and beyond. But this new tradition does not mean that the Scientific Revolution has no connection to the past. Indeed, we should all agree that it began the scientific questioning of the world based on a new systematic procedure of observation, hypothesis, and validation. The Scientific Revolution grew out of medieval science that changed in the sixteenth century. It was not based on reputed authorities that offered an interpretation of science, such as Aristotle, Ptolemy, or the Bible, but on experimentation and observation that prepared the way for an inductive approach to knowledge. It began a new way of assessing science that was concerned primarily with physical and mathematical evaluations of matter. It also benefited from the cosmological interpretations of Neoplatonism that questioned the views of Aristotle and the concept of a finite universe. Although the views of Copernicus were attacked and ridiculed for many years, even by Catholic and Protestant theologians, eventually they were accepted as true, or at least, were accepted as indicating where astronomical truth could be found. Only in the seventeenth century did the Catholic Church commit itself to opposing the interpretation of Copernicus, a commitment that locked the church into a position that would prove to be nearly impossible to defend.

It would be a small matter if the only effect of the Scientific Revolution was a challenge to astronomy, but it also challenged the idea that humanity was the sole reason for creation. Given that the world is seen from a purely human perspective, it is understandable that everything is presumed to revolve around us. And therefore any doubt cast on humanity's favored position may be considered to be a bold attempt to dispute this claim. In fact, there are two ideas contained herein, each of which reinforces the other. One idea concerns anthropocentrism or a human-centered universe that is accepted as true by the Bible, particularly, in the book of Genesis. The other idea concerns the Christian interpretation that the first idea is confirmed by the workings of the universe through the creation of the world, and the birth of Jesus Christ as a necessity for salvation. Hence, anthropocentrism or a human-centered universe reinforces a Christian interpretation of science, and a Christian interpretation of science presupposes anthropocentrism.

The practical effect of these two ideas concerns a matter of grave importance. Since it was understood that the existence of other stars, probably numbering in the billions, if not trillions, meant the likelihood of other planets that in turn inferred the possibility of other intelligent beings, then humanity's favored position that originated with Adam, followed by his disobedience of God, original sin, and ultimately humanity's salvation through the birth of Jesus Christ could no longer be true because the appearance of Adam and the unique birth of Jesus were thought to occur only once on this planet never to be repeated elsewhere. If these other planets were inhabited (and they should be inhabited because God would extend his goodness throughout the universe), then Jesus would have to make an appearance there also, which would destroy his unique occurrence on the earth. It would also undermine the purpose of the Catholic Church that continued the ministry of Jesus after his ascension to heaven. Therefore, the greatest effect of the Scientific Revolution was not on science, but on metaphysics, and this effect indicates why it plays such an important role in the development of secularization. If a Christian believer followed the consequences of the Scientific Revolution to its logical conclusion, he should have grave doubts about his religious beliefs. A believer could still make a leap of faith, but it would be over a great chasm.

Contemporaneous with the Scientific Revolution was the Protestant Reformation, which described the religious transformation of Europe between the Renaissance and the Enlightenment.[6] Although the Catholic Church attempted to retain its privileges that it had acquired over several centuries, the corruption of late medieval and Renaissance popes, who acted more like temporal rulers rather than spiritual leaders, defined the background to the Reformation. The Protestant Reformation was disruptive to European society, but it also revived Christianity by producing several new sects and denominations. And this revival occurred with the substitution of the Bible (as an authority for Protestants) for the Catholic Church (as an authority for Catholics). Despite the denial of the necessity of the Catholic Church, the Protestant Reformation fostered a religious inwardness that promoted Christology, that is, it promoted a reverence of Jesus Christ not as recompense for a heavenly reward, but because Jesus, as God, was worthy of devotion, regardless of any otherworldly consequences. By its very essence, Protestantism privatized religion. Certainly, by the middle of the seventeenth century, most European states had been secularized, that is, they enforced the separation of church and state, and pursued worldly goals as their primary concern. These actions did not deny the existence of God, but they did reduce the influence of religion on society. It is within this context that the Enlightenment emphasized the use of reason when dealing with the world, and questioned the usefulness (and power) of the Catholic Church.

In conjunction with the Scientific Revolution and the Reformation, the philosophical system of René Descartes (d. 1650), known as Cartesianism, developed a rational interpretation of reality. It was dependent on *a priori* or deductive reasoning as the basis for this rational interpretation. Knowledge would be subjected to a universal truth that is derived from self-reflection, that is, one's own self-awareness became the basis for understanding truth (rendered succinctly in the expression "I think, therefore I am," or in Descartes' Latin version, *Cogito, ergo sum*). One's own thought (the self) and the reality of everything else (the material world) were regarded to be independent and disconnected from each other. Therefore, mind and matter are distinct realities. Matter is evident in objects (or bodies) and it shares a characteristic in common that Descartes called extension. Since an object or body exists because it is an entity that is "extended" in its shape, size, motion, and other physical features, Descartes denied extension to anything that is not physical. Hence, matter must be distinguished from thought because the latter is associated with the mind only.[7]

Cartesianism not only promoted an intense reflection about reality, it also emphasized that whatever could be known was what could be understood. Therefore, nature was reduced to physical phenomena, to mere matter and motion. This line of thought did not deny the existence of God. Because Descartes presupposed that an infinite God exists, he also presupposed that we could not understand anything beyond human comprehension, and both God and infinity would be included in this distinction.[8]

Although it may be said that Cartesianism had little or no influence on non-scientists and non-philosophers, it did have an influence on the development of secularization in European culture because it was perceived to be a threat to the authority of the Catholic Church. Because of this threat, Cartesianism was attacked for two main reasons. It was attacked because it challenged scholasticism that had preserved Aristotelianism, particularly, its view that the earth was the center of a finite universe. Cartesianism was also attacked because it allegedly led to freethinking that criticized the teachings of the Catholic Church.

TECHNOLOGY'S ESSENTIAL CHARACTERISTICS[9]

Assuming that there is a meaning to be found, the evolution of secularization since the Middle Ages has made the present age an unfortunate time because it has made the quest for meaning ambivalent. It has undermined the struggle between time and timelessness, between living in a temporal world while preparing for an eternal afterlife.[10] Because answers to the questions of the ultimate meaning of life are hard to find, some of us persist in our religious beliefs, or follow fundamentalism that is alien to modernity, while others believe in nothing and bring nothing into the world. A secular age may encourage people to idle away their lives with frivolous activities and diversions, who devote themselves to their jobs or careers because they have no other devotion. When removed from the customary routines of life and when given the opportunity to do so, many people seek the fulfillment of themselves through work because work is tied to the quantitative conditions from which it is derived. Because we are all very much preoccupied with activity, none of us really has time for anything; by being busy, it is difficult to reflect upon the world or ourselves. A secular age deifies work because there is so much of it, but the pace of our age leaves little time for introspection. To some extent the present age has replaced spirituality with occupation, thereby making work (with or without a work ethic) a substitution for prayer. Unlike the preoccupation of religion in a theocratic age, the present age has made a choice between belief in something beyond itself and belief in something in spite of itself.

But this is to be expected. It is expected that a secular age would pull away from religious feelings, at least from its public manifestations. It is expected that the two principal problems encountered by religion—explaining the inexplicable, and overcoming or at least justifying the fear of death—can only now be confronted in a different way. When true and complete answers are lacking to the questions that afflict us, we refuse to seek them out, and instead opt for fictitious ones. Although religion reaches out beyond ordinary human experience by tapping into the unlimited resources of a divine presence that is believed to transcend all knowledge, intellectually it introduces us to platitudes of meaningless rhetoric offered to give consolation to those who have a need for such limitations. Lest we forget, however, secularization weakens religion on the one hand because it challenges tradition while stimulating new religions on the other because it creates the opportunity for new beginnings.

Even when derived from personal choice, social obligations tie religion to society. The religiosity of a particular society will differ from other societies, that is, a Buddhist society will be affected differently from a Christian society as the latter will be affected differently from a Muslim society. Even within a society, religious sects will be affected differently, as a Catholic society will be affected differently from a Protestant society. Similarly, technology is expressed differently by individual societies, depending, in part, on the social obligations inherent within them, that is, the social obligations that relate to technology help to define the latter's reception by society. We may conclude, therefore, that social obligations tie both religion and technology to society.[11]

Although presented with an opportunity, technology cannot answer the questions of life and death, being and nothingness, God and the cosmos (unless they are redefined), but this inadequacy does not prevent technology from being molded into a religious form. Technology is not prevented from devising a way of life, nor is it prevented from demanding obedience to its commands. When stimulated by secularization, technology advocates the coming of a certain type of kingdom, stressing the beneficial value of progress and the righteousness of the commandment of scientific inquiry. It demands renunciation of old religions as wrong and acceptance of itself as right. It categorizes nonscientific writings as apocrypha. It demands recognition of itself not only as a rite, but also as a way to a theology, in a word, as a faith.

By accepting technology as a religion, technology challenges other religions because it repudiates customs offensive to itself. Technology can easily become a religion because it already possesses religious characteristics, including proselytism. For example, its rites of passage are evident in its techniques, its priesthood in scientists, technicians and engineers, its temples in the academies of science and university laboratories, its catechisms in technical journals and manuals, its doctrines in scientific rationality, its ecclesiastical councils in professional conferences, and its sacred history in the idea of progress. The only thing that technology lacks is a prophet who can interpret its meaning or express its will. As a religion, technology has a long history, or to be more precise, a long prehistory, that reveals humanity's invention of technology, and its necessity to human survival. It is comprised of procedures and methods, as well as products and devices. It is humanity's most comprehensive creation, spanning the entire time we have been in existence.

As a universal truth and regardless of the age in which it is used, technology is oriented toward the future. To believers who profess a faith in technology as the only truth, its priesthood, that is, its technicians, has become ecclesiastical authorities, authoritative interpreters of sacred texts, who encourage believers of other faiths to apostatize. These priests stand ready, like all priests, to administer to that one important aspect characteristic of religion, to a conviction thought to be greater than the self. When we believe in religion, we question ourselves. We accept it without criticism, without standing back and looking at it honestly. It may give us assurances of certainty when times are uncertain, and the need for certainty may be most acute during economic or political upheavals. A struggle against religion is a struggle against someone else's conception of the world, like the mode of insight proposed by Hegel's obsessive system in which everything is meant to fit with utmost precision. Although the categories of gods, spirits, and ancestors may be self-generated,[12] religion still receives our highest respect. Much like religion in general, we do not alter technology's doctrines, for even when threatened, technology is applied aggressively. Another similarity technology has with religion is its portrayal as the means to salvation, as the way by which we can be saved from our own shortcomings and saved for our own well-being.

TECHNOLOGY'S FATALISM

Although religion is not confined to a single, individualized idea, but to a sum greater than the parts that go to make it up, it is always tied to an organization that is followed by varied and intricate beliefs. By comparison, it may be observed that technology itself is a poor example of religion because it lacks this plurality of beliefs. Secular and monolithic, technology contains no magic and no mystery, and lacks the notion that religion is based on sacred writings, divinely inspired. Nor is religion, unlike technology, reducible to one idea, because each individual religion consists of a plurality of ideas, each interlocked with all the others. Of course, when religious beliefs are limited in any way, the religion of those beliefs disappears. It is not so much that technology is opposed to religion; rather, it stands against religious speculation and wonder, against revelation without dependence on immutable laws and irrefragable truths, against a divine inexplicable being who is unattainable and defies comprehension, sitting on a throne, flanked by its supporters. Technology introduces its own notion of truth: worldly, one-dimensional, and precise.

Religion is not displaced by technology because religion is an institution well-endowed with bureaucracy that itself possesses its own techniques. Because the truth of religion is questioned in a secular age, it is no longer believed to be the only path to divine grace. As we mentioned above, secularization has little influence on religious organizations already established, but much influence on their respective societies. The more technology becomes esoteric, the more it removes itself from the masses, particularly what the masses regard to be truly meaningful, even if such beliefs appear base and banal to the sensitive. Therefore, religion, which is now less literal, has little to fear from such rivals, even when science, which is now less materialistic, seems to deprive religion of its usual explanations of the nature of the universe and the origins of humanity. What it has to fear instead is a degenerating indifference, sustained by boredom, despair, and disenchantment, whose only redemption is salvation at the hands of as pharisaic ministry.

It cannot be denied that we have passed through an age of transition. We should not ignore the importance of history on secularization, but we should also not ignore that secularization may usher in new religions. A secular age proposes the question: What new religions will appear? Although unknown, one possibility is technology. One must admit that there is some irony in this idea, that although traditional religions have been weakened, the notion of religion is now associated with the essence of technology. Despite its historical roots, this transformation has come about because technology elevates its own view of reality—while simultaneously suppressing impulse and instinct—into a matter of religious experience. This tendency becomes evident the more technology evolves. Like religion in general, technology is both spiritualistic and moralistic. Although there is a relationship between the two, religion is not the sole basis of morality.[13] Locke's arguments that religion and morality are related were never true, but at least moralities do have a basis in fact, that is, they have validity because they concern how people live and interact with other people. Despite these observations and despite technology's attempt to alter morality, technology remains an inferior type of religion. Because religion gives feelings of dependence in the sense defined by Schleiermacher,[14] it is a dependence relating to belief. Technology also fosters dependence, but its dependence is tied not to a hidden realm of being, but to overtness, to a dependence that renders technology into a dispassionate, insistent, and systematic religion.

Technology calls into question religious forms by challenging those categories of thought that we are most familiar with and that appeal to the most basic feelings even when they no longer suffice. The present age questions the relevance of all religions, and in the West, the relevance of Christianity. Although the rigors of Christianity have been weakened, no matter how much reverence they may still be given, technology would not have advanced as far as it has without the profound impact of Christianity. To put it simply, technology is the child of Christianity, and a scientific, Christian Europe. To which it may be added that technology's past experience goes a long way to explain why it takes such a linear view of the world. It explains why it lends itself very easily to exploiting and dominating nature, that is, why it proselytizes with sword and book. Hopefully, we will go beyond an obsession with a once transcendent deity who has allegedly revealed himself to humanity. Although weakened, Christianity's metaphysical aspects may survive, the metaphysics that depicts an individual, apprehensive self-doubt before the absolute, of a solitary self before the infinity of being.

Nevertheless, the waning of religion ushers in technology's own fatalism. One myth of technology is the notion that it can absolutely determine the course of the future, that everything can be planned, controlled, and regulated to the smallest degree. This statement is one description of "the technological imperative." If we idealize technology, it is an ideal we can know and understand. But such a notion is opposite to that which says that we are predisposed toward a belief in the unseen and unknowable, a belief that corresponds to the Greek idea of fate. This is why in the last resort technology is a poor example of religion. It strives to erase life's mystery, and thereby gives us nothing after we accept it, except more of the same. Although religion has created a transcendental realm to the abandonment of everything else, technology in its opposite mode is no better. Religion has advocated that because our bodies confine us, we must let our spirits soar. Yet neither religion nor technology refuse to accept humanity uncontaminated or untainted as a legitimate realm of influence, both of which manipulate and exploit humans because both distrust them. Both regard humanity as morally impure, repulsive, and contemptible, and both try to make humanity clean, simple, and obedient.

While technology is rational and voluntary, religion is irrational and involuntary. Accordingly, the excessive rationality of technology must be contrasted with the arbitrariness of religion. Naturally, there are fundamental differences between religion, which looks to the past, and technology, which theoretically at least, looks to the future. Religion mimics what has passed away because it concerns the attainment of a life that has already taken place. This worship of the past is called tradition, and tradition is merely a longing, not a way of life. It is little more than sentimentality, since it is an admiration for what has been. Technology, on the other hand, looks to the future, and Christianity's influence on technology has intensified this tendency. Although religion is only one type of experience common to humanity, it is closer to social truths (the truths that relate to others) than to ontological truths (the truths of the self). When religion is influenced by philosophy, it makes an attempt at least to acquire a basis for ontological truths. But replacing religious values with technological ones has not eliminated the need for convictions because just as a person of faith is tied to his convictions, and therefore finds it difficult to make concessions, so too is the technician.

One effect of secularization on technology conforms with the description of religion as a means to power. But in order to be powerful, power needs continual reassertion. Without the latter, power dissipates. Power, and most notably technology's form of power, is a means to an end, but never an end in itself. Not only may technology evolve into a religion of power, it also manifests the power of religion. Since power is conditional, it is limited to and defined by circumstance. Changing situations automatically change the conditions of power. This is another reason why technology needs stability because the latter confirms the maintenance of control. Nevertheless, we now encounter a contradiction because technology in no way can introduce real, substantive changes into the world, changes that might so modify the world as to be a threat to our existence. The purpose of technology is to aid human existence, not to hinder it. Like religion, we give to technology what we deny to ourselves. When technology shares its attributes with us, we become empowered. Although technology can make us externally powerful, it ignores us internally. Ultimately, technology merely alters the shape of things to come, but does not change them, since to alter anything means to make things different without changing them essentially into something else. Genuine change can only originate with us, and it is us who make technology powerful.

TECHNOLOGY'S MEDIOCRITY

Even without intending to do so, secularization has heightened technology's omnipotence and omnipresence. It has decreased the relevance of established religions, while increasing technology's own relevance. As a result, technology may be capable of dictating the meaning of piety. But the shortcomings of technology as a potential religious form are most evident in its application by the masses. Since the masses are rendered possible because of technology, technology creates the situations that aid their growth. Despite the seemingly favorable conditions of better health and increased production, technology also engenders apathy and indifference. But the characteristic that most accurately epitomizes the masses, and is most characteristic of technology as a religious form is mediocrity, which is a chief characteristic of Western technology's Christian foundation. Mediocrity is what makes technology so appealing to the masses because it makes everything possible. Every hope, every desire, and every wish can come within the reach of everyone. But what is possible is what is feasible, and what is feasible are the small things. The attainment of equality, which is so much a preoccupation of our time, ultimately gives the masses very little to be proud of because they would have achieved an equality with what everyone else has achieved, but not what they strived for, even if they have failed to achieve success.[15] Running in a marathon, for example, is no longer a race characterized by speed, but by participation. For many participants, winning the race is no longer the issue, but being there is. It is similarities, not differences, which unite people. Consequently, technology as a religion promotes mediocrity, and as a result, it lowers the prospects for humanity.

Since the masses embrace as well as encourage the ideal of mediocrity, they deny individuality, the promise of which is one of technology's great illusions. Just as individuality was suppressed by many religions, particularly in the axial period as described by Jaspers,[16] technology does the same. In fact, the greater the mass, the closer it approaches the lowest common denominator, and therefore, the more mediocre its character. This it cannot prevent. Technology cannot forestall the natural occurrence of mediocrity at a time when we have become so numerous. The existence of the masses not only gives its participants a false identity; it also allows them to doubt their true selves because the leveling process of mass culture is expressed as a denial of the self. It has become a form of self-hatred.

And to acclaim the masses means eventually to obliterate the possibility of heroes. Heroes transcend the masses. Although they originate in the mass, they rise above the limitations imposed by mere numbers. Since there is nothing extraordinary about a multitude, it is hardly the place to find heroes. Religions do not contain heroes, nor do they contain hero worship. The nature of religion promotes mediocrity, at least in its initial intent. The closest religion could come to heroes are saints who achieve notoriety through mortification, shame, and humiliation. Although possessing a consciousness posed against itself, no religion contains truly heroic behavior.

It is also because we have no genuine deities who speak for us that we have no heroes; for wherever we have heroes capable of greatness we have gods with the same tendency—both of which are capable of performing mighty feats, just as each dares to defy the other. In place of heroes, our age has hero substitutes, that is, individuals superficially heroic, such as actors, athletes, and popular entertainers who are really surrogate heroes, whose popularity is sustained by their appeal to a mass audience, although their form of heroism is always peripheral. These modern heroes are idealized by the masses because they are believed capable of accomplishing wonders beyond the reach of the ordinary. Hence, they are idolized. But as we said above, mediocrity and equality deny the hero. When genuine heroes exist, the situations which create them are far more discriminating, to say nothing about the idea that heroes themselves distinguish between alternatives. Yet the greatness of the hero rests on more than what he does. It rests on who he is, and who he is plays a part in what he does. To be at the right place at the right time is no more a qualification of a hero than luck is. No matter what else may be unnecessary to heroism, talent, that is, excellence, always is necessary.

Hence, true heroes are particular, and all particularisms are anathema to technology because technology is the great equalizer. Like democracy, technology innately distrusts heroes because the latter conflict with the former's mandate for equalization. Because technology loathes differences between distinct entities, the lazy and marginally successful are paraded with the energetic and gifted so that not so much are the lowly raised as the exceptional debased. In this connection, there is a reason why technology brings absolute decadence because it consists in the elimination of creative energy.[17] It deprives us of the source of originality, which has been severed from us, like a form of castration. Of course, if you are equal, you are interchangeable, and if you are interchangeable, you are expendable. Equality does not liberate the individual, but obliterates him. Nor does equality increase liberties because it shrinks the differences between the exceptional and the ordinary. If there were any laws of averages that regulated human affairs before the advent of our age, they were certainly reinforced after the dawn of it for the kind of religion manifested by technology is truly for everyone. Even if technology fails ultimately to bring equality, at least it brings the demand for it.

CONCLUSION

Of course, technology does not propose final answers to the questions it asks. It is indeterminable. Religion, on the other hand, proposes determinate and obtainable ends for us. Whereas technology seeks to gain more knowledge within the infinity of knowledge, existing religions cast a pall on technological pursuits, particularly, when we consider the relationship of knowledge to technology and religion. Technology, now accompanied by science, is the pursuit of the knowable, but religion is the worship of the unknowable. One effect of technology aids science to uncover and to probe the unknowable so that it may be knowable. The purpose of religion is to accept the unknowable for what it is and to leave it concealed because the nature of divinity is hidden. It is assumed that divine intervention, that is, revelation, is necessary so that deities can make their intentions known. Religion ultimately is concerned with the unpredictable, the unintelligible, the denial of what is at hand, and the affirmation of what is not. As astutely delineated by Kierkegaard, we must make a leap of faith in order to believe in any divine entity.

But technology has changed this idea. It has changed the concept of religion because it attempts to reduce speculation and doubt to fact, to replace wonder with certainty. Technology attempts to erect a wall of scientific facts as a defense against revelation. It also affects the development of new myths for old ones. To say that all religions make use of some type of technology is not to say that this use will evolve into a religion. The use of technology is more sophisticated and more involved than the lighting of a candle or the swinging of a prayer wheel. It entails a total involvement in the elaborate structure of a new paradigm. It substitutes a new approach to life for an old one.

Since religion can become irrelevant by perpetuating meaningless banalities, particularly, by dispensing assurances of some distant afterlife that will go on everlastingly, it can become a wasteland. It lacks both leadership and real practicality, and is far from understanding being. But, then again, religion easily falls prey to technology's mechanization that challenges and demythologizes religious mystery. Modernity in general (that includes the influence of secularization) and technology in particular have delatinized religion, making it as monolithic and as sterile as mechanized agriculture. Technology makes all religions useful, that is, it makes them practical, but ultimately it renders them useless. Thus, Nietzsche's "death of God" encompasses a duality; it is both a liberation and a deprivation. If we can come up with something better, it is a liberation, but if we cannot, it is a deprivation. Religion may still linger, but if we fail to devise new religions, the old religions will be much more attractive—a pity, since they offer us little real consolation. It is people who fail and then succeed, who sin and then repent, who err and then reform. The gods wielding thunder bolts, appearing as apparitions after death, or speaking through the clouds are poor substitutes for the totality of being and the challenges of life.

When compared with the past, the present age repeatedly finds itself at a loss. Although secularization has made religious values weaker, their unverifiable prophecies still persist among us, giving cause for hesitation. We may follow the old religions or we may follow none. The elaborate structures of these religions are reduced to a simple and poorly defined belief in something remote, but we continue to be at a loss even when we reflect that technology itself has a religious calling. Regardless how we perceive the present situation, we live in uncertain times. The loss of spirituality in the true and profound sense of the word is a great detriment to us. Such a loss has deprived us of the self-appointed assurance that comes with a transcendent realm of being. As a result, we have been condemned to live in a world of objects, and like Adam expelled from Eden, we feel homeless in the world. Destitute and desperate, we make false idols and listen to false prophets.

Although there is a need for a new spirituality, hitherto undefined, it is of little demand in the world, a spirituality that can define both the beauty of humanity and the wonder of being, a spirituality that goes beyond any historical god, but does not ignore humanity. The principal accusation we can make against secularization and the latter's influence on technology is that technology as a religious phenomenon develops out of a perversion of what is truly meaningful. Since religion has survived and prospered primarily because of a decline in freedom, it imposes a preconceived perception on the world from which responses are expected, responses indicating that religion is primarily a belief system tied to a following. Like any religion, technology manifests specific and well-delineated attributes, since they are formatted for a matrix.

NOTES

1. For signatures and their relationship to secularization, see Giorgio Agamben, *The Signature of All Things: On Method*, trans. Luca D'Isanto with Kevin Attell, New York: Zone Books, 2009, pp. 76-77.

2. For an introduction to social differentiation, see Talcott Parsons, *The Social System*, Glencoe: Free Press, 1951, pp. 201-207 and 362-364.

3. Secularization is a well-discussed topic in sociology of religion. In reference to the idea that secularization has ceased or is a false theory, see David Martin, *On Secularization: Towards a Revised General Theory*, Aldershot: Ashgate Publishing, 2005, pp. 17-25.

4. *The Defensor Pacis of Marsilius of Padua*, ed. C.W. Previté-Orton, Cambridge: [Cambridge] University Press, 1928, dictio II, capitulum 4 and 21. Also see *Marsilius of Padua, The Defender of the Peace*, Translation and Introduction by Alan Gewirth, 2 vols., New York: Columbia University Press, 1951-1956, II (1956), pp. 113-126 and 287-298. The *Defensor Pacis* was completed in 1324.

5. Thomas S. Kuhn, *The Copernican Revolution: Planetary Astronomy in the Development of Western Thought*, Cambridge [Mass.]: Harvard University Press, 1957, pp. 192-193. Copernicus' book, *De Revolutionibus Orbium Coelestium*, was published in 1543, the year of Copernicus' death, and was later placed on the Index of Forbidden Books by the Catholic Church in 1616. It was removed in 1835.

6. As part of the religious transformation of Europe, mention should be made of the printing press that was invented in Europe in the middle of the fifteenth century. Although it was an aid to the study of the Bible, it was also a challenge to the literate monopoly of the Catholic Church. It allowed scientists to disseminate their discoveries that in turn may have helped with secularization. See Elizabeth L. Eisenstein, *The Printing Press as an Agent of Change: Communications and cultural transformations in early Modern Europe*, Volumes I and II, Cambridge: Cambridge University Press, 1980, pp. 520-543.

7. René Descartes, *Principles of Philosophy*, pt. I, sect. 53 and pt. II, sect. 4. John Cottingham, *et al.*, trans., *The Philosophical Writings of Descartes*, 3 vols., Cambridge: Cambridge University Press, 1985-1991, I (1985), pp. 210-211 and 224. Since the time of Descartes, science is more likely to substitute the brain for the mind because the brain can be identified. The mind, on the other hand, means human consciousness that originates in the brain.

8. Descartes, *Meditations on First Philosophy*, third meditation; reply to the *First Set of Objections*; and reply to the tenth objection to the *Third Set of Objections*. *Ibid.*, III (1991), pp. 32, 81 and 132.

9. Material for this section and the following sections in this chapter are based on chapter 7 ("Lost Among the Stars: The Secularization of Religion") from my *Contra Technologiam: The Crisis of Value in a Technological Age*, Lanham: University Press of America, 1993, pp. 83-92.

10. Richard K. Fenn, *Time Exposure: The Personal Experience of Time in Secular Societies*, New York: Oxford University Press, 2001, p. 6.

11. There is some doubt if modern science has led to the decline of religiosity in the non-Western world, but it certainly is relevant in the Western world. If secularization occurs in the non-Western world, it is caused by Western values challenging the way of life that is evident there. Since religious beliefs have social obligations, social obligations impact religious beliefs. One notable example concerns the role and treatment of women. In societies where women have limited opportunities, women are involved primarily with marriage and children, which generally emphasize the importance of religion. In societies where women have more opportunities, women may

be employed outside the home, which may deemphasize the importance of religion. Therefore, the role of women and women's rights may aid secularization, regardless where it appears in the world.

12. This idea is the predominant theme in Pascal Boyer, *Religion Explained: The human instincts that fashion gods, spirits and ancestors*, London: Heinemann, 2001.

13. John Locke, *An Essay Concerning Human Understanding*, ed. Alexander Campbell Fraser, 2 vols., Oxford: Oxford University Press, 1894, I, p. 474 (bk. II, ch. 28, sect. 5).

14. Friedrich Schleiermacher, *The Christian Faith*, 2nd Ger. ed., trans. H.R. Mackintosh and J.S. Stewart, Edinburgh: T.T. Clark, 1928, pp. 12-18.

15. It is true, we should say, that our greatest strength is not avoiding failure, but attempting to try again every time we fail.

16. Karl Jaspers, *The Origin and Goal of History*, trans. Michael Bullock, New Haven: Yale University Press, 1953, pp. 3-6.

17. José Ortega y Gasset, *The Revolt of the Masses*, anonymous English trans., New York: W.W. Norton, 1932, p. 43.

Chapter Four

Technology and Religion

A Metaphysical Challenge

INTRODUCTION

Religion serves many purposes and addresses many needs. It proposes a code of conduct and helps to bind people to a particular culture. It deals with a system of rituals through which one must be initiated and offers an explanation of the meaning of life. It reflects upon the human condition and confers a sense of immortality. It is built on an objective order to the world beyond human experience, an assumption that may be the fulfillment of some need we have, regardless of reality. Despite the form religion takes, it attempts to answer basic questions: Who am I? What is my purpose in this life? What made everything that exists? But an evaluative examination of any religion asks other questions: How much is religion a part of culture? How much can religion be changed and its respective culture remains the same? How much does religion accurately describe the world and its relationship to being?

It is evident that religion is an essential characteristic of human culture and lies at its foundation, but how religion originates is not the result of critical thinking. Religion has been defined as a belief in supernatural forces, but this definition has been variously interpreted. This belief may be tied to some supersensory place where one's soul will reside after death in the absence of any god (as in Buddhism), or to a supersensory place where a supreme "being" that is thought to be all-powerful and all-knowing resides (as in Christianity or Islam). Religion may be defined as a belief that addresses the ultimate questions of life, but this belief has little to do with philosophy, although the latter also addresses the same questions. A philosophical analysis of reality should not be confused with a religious analysis. Religion may be defined as the result of its function, that is, it may be defined in relationship to its role in binding society with its people. We should remember that the word religion is derived in all likelihood from the Latin verb *religare*, which means to bind or to tie.

Although attributed to a foundation based on reason, religion remains uncritical because it contains no provable propositions. It offers no plausible explanation for the world.[1] Rather, religion is thought to be the result of some sort of revelation that presumably demonstrates a causal connection between the world and the manner in which it might have come into being, or why it functions the way it does.[2] It is based on preconceptions that the world can only be understood when supernatural forces are introduced, even though there may be no actual causality between them. As a result, religion depicts a world different from the world we live in, and after depicting this other world, it then attempts to explain the world we do live in. Since each religion offers its own interpretation of the world, each religion is different, despite minor similarities. And yet religions are bound to culture because they participate in a common human experience that uses rites, rituals, myths, and taboos in order to be understood.

There are many historical examples in which different religions have proposed explanations of this alleged causality, a few of which are described below. The Nile river, for example, was vital to the survival of the ancient Egyptians, and if the water level of the Nile was low, this event occurred because the Egyptians believed the god Osiris, who they associated with the Nile, was angry with them and thereby showed his displeasure. Or Christians and Jews say that unless one abandons an adulterous life, God will punish the adulterer, and this occurrence will happen despite other biblical references to the contrary. Or when the city of Rome was sacked by Alaric and the Visigoths in the early fifth century because the Christian Roman emperor Theodosius I had closed the pagan temples and had prohibited the worship of the gods that were believed to offer protection against invasion. In these examples, there is an assumed causal connection between the level of the Nile river and Osiris' approval, the assumed connection between adultery and God's threat of punishment, as well as a connection between the sack of Rome and the closing of pagan temples.

Although there is much in common between goals of theologians and metaphysicians, we need to ask if religion is a substitute for metaphysics. Is a religious belief a distortion of metaphysics, that is, is it a perversion of the analysis of being? When we talk of religion, do we mean the plenitude of being, the power of creation and the depth of reality? What was the sun god among the Sumerians, Egyptians, or Mayans, if not an attempt to understand the importance of the sun in relationship to the world? Who were Buddha or Jesus, if not conscientious moralists who attempted to address the pressing issues of their day? Who are the so-called supernatural gods, notably, Yahweh and Allah, if not attempts to find an ontological explanation for reality by an absolute being that allegedly transcends it? More than the religious explanations that have been proposed in the past that have given guidance to many people, there remains the wonder and depth of being.

If we wish to understand the plenitude of being, we must become aware of its omnipresence. Hence, we must ask where being is to be found. It is in you, in the world, in fact, in everything in the universe, manifesting itself in all its majesty.[3] Religion may be conceived as an attempt to understand ourselves and the world in relation to the forces beyond human control, but science gives us a more plausible explanation. For example, we do not have to believe that the earth is a sphere; rather, we know it is. Religion, on the other hand, cannot give us assurances of its beliefs because they are not based on fact, that is, on the reality we know and experience everyday. And what is philosophy, especially metaphysics, if not an attempt to make sense of the world through reason and analysis, nor should religion be confused with spirituality, the belief that being is uninfluenced by any physical or material presence. Schopenhauer defined religion as the way in which people comprehend being. Therefore, religion serves as (and becomes a substitute for) metaphysics.[4] As comforting as any religion appears to be, it is not grounded in reality because it assumes there is something beyond it, but the same accusation inaccurately has been made of metaphysics.

THE THREAT OF TECHNOLOGY

In light of this evidence, technology has become a most powerful force because it challenges religion wherever it appears, but at the same time, it lays the foundation for itself as a religion. Technology may be defined as the practical applications of human creativity. It includes the effects of science, but it also includes material and rational interventions with the world. It embodies both the alphabetization of language and the mathematization of numbering systems. Nevertheless, technology does not deliberately challenge any religion because it contains no mechanism to do this. Rather, technology challenges religion because of its accumulative presence in the world. It is not an opponent to any religion *per se*, and yet religions are increasingly challenged by technology whose presence is everywhere. Any confusion between religion and technology lies with religion. Although traditional religions have lingered into the present age, their believers follow their own traditions with little reflection. Those who attend religious services in churches, synagogues, mosques, or temples seem to be unaware or indifferent to how the world has changed.

Although technology does not possess a religious mechanism, it does possess characteristics that can be used religiously, and this is why technology, in addition to its historical causes, can become a religion. The more technology has evolved, the more it perfects these characteristics, but technology as a religion should not be confused with technicism that may be defined as the reverence for or the worship of technology. Although these characteristics will be discussed below, suffice it to say here that these characteristics are innate to technology as it has evolved, characteristics that have helped technology to perfect its application in the world. Therefore, on the one hand, technology threatens religion, and on the other hand, it may evolve into a religion itself. The old gods of the sun, burning bushes, thunderous voices, crucified reformers, traveling preachers, residents from the highest mountains and the deepest seas will be followed in time by yet another creation that addresses and fulfills our yearnings. Whatever failings religion in general has, technology may also possess.

Despite the possibility of becoming a religion, technology may offer only moderate improvements over existing religions. It may offer little consolation in comforting the weary, giving hope to the forlorn, or conferring mercy to the guilty. These shortcomings are due not to inherent imperfections in technology, but to our expectations. We expect too much from technology and assume that it will satisfy all our needs. The old religions may also have been imperfect, but at least they were more forgiving of what they could not do. We believed in them without any serious doubt. We prayed, or lit a candle, or bought an indulgence, or made an offering, or were contrite when we confessed our sins, and we did these things never really knowing that they had an effect. We believed in our belief. Yet, it is no irony that technology may perpetuate religion into the future, but it will not be one of its best forms. So the question of religion and its association with metaphysics, as discussed above, will remain even after technology becomes a religion.

Since religion in general is thought to embody a meaningful morality, what chance does technology have in attaining one? How can technology give comfort to the tribulations of daily existence? If we were able to define the meaning of existence, would we reach a consensus? Once we remove politics, or economics, or any other limitation, and we look at reality unencumbered, what do we see? Do we need some superimposed mechanism in order to function in society (as the Aztecs practiced human sacrifice to assure that the sun would rise in the morning), or can we live without the need of the supernatural? An answer to these questions is imperative because a meaningful morality is directly related to a helpful understanding of life, or so we assume. If technology is able to help us understand who we are and where we are in the world, if it can give us comfort and help us live meaningful lives, if it can close the gap between a feeling of purposelessness and an understanding of the wonders of reality, if it can cure disease and promote life, then it has a chance of developing a meaningful morality. Technology can do these things when it is free from hypothetical conceptions of what reality should be. As soon as we presuppose that there is a world beyond this world, that is, a spiritual realm of heaven and hell, we lay ourselves open to all kinds of "persuasions." Therefore, if technology is limited in some way, the perfection of a meaningful morality is hampered. In order to be meaningful, technology must be free from traditional morality.

RELIGIOUS APPLICATIONS

Although religion is expressive of its age, it can also linger in an age that no longer supports its meaning. Its believers can prolong it because of tradition or habit as it can survive in the absence of a meaningful conception. It can survive in the backwaters of a culture. It may even fall back into fundamentalism that breathes intolerance and a fear of modernity. When threatened, religion may also undergo adaptation. Regardless of shifting circumstances, religion may attempt to modify its beliefs only to such an extent that its fundamentals are not denied. It may even ignore some of the rituals that reinforce its theology. Religion can adapt to changing circumstances because that is the easiest way to avoid confrontation. How many times has religion accustomed itself to circumstances that formerly were opposed to its own beliefs? How many times has religion survived because its followers went underground? Hence, we can say that religion may become anachronistic.

Are traditional religions out of place in a technological age that is characterized by a secular or non-religious conception of reality? How can Islam's primitive tribalism, Buddhism's guaranteed route to nirvana, Judaism's immobile ritualism, Taoism's all-inclusive naturalism, Hinduism's elaborate hierarchy of gods, Catholicism's residual medievalism in form and substance, or Protestantism's adherence to strict biblical interpretation as the only way to salvation find meaning in the present age? How long will these religions persist in an age that has surpassed them? How long will people boldly display a religious symbol hanging from a chain around their necks as if it represents something real?

In addition to its alignment with metaphysics, religion has also been used for other ends. It has been used, for example, to incite people either to invade a region or to defend one's own, to persecute a sect or group of people disliked by society, to give encouragement to people so they may achieve a higher cultural or ethnic well-being, to reinforce the assumption that the past may have been superior to the present, or to justify many types of political activity, even those that may be questionable, if not reprehensible. In these examples in which religion serves additional purposes, can technology do the same? Can technology be used to defend or invade a region, to be utilized internally when one faction of society opposes another, to encourage the betterment of a people, to show that the past may really have been inferior to the present, or to be applied to all types of political activities, even those shameful? Who would deny that these things can happen? Who would deny that technology and its concomitant religiosity can be formed and utilized like the beliefs and doctrines of any religion?

Since religions are arbitrary, we should respect the religions of other people. If we follow a particular religion, it was probably the religion of our parents, and of their parents, and so forth back in time. Because religions originate at some point in time, there are often problems concerning the validity of their origins. Do religions originate from something of great importance to humans (Feuerbach), or from the promotion of an existing condition or maintenance of the status quo (Marx), or as an aid to life itself (Durkheim), or from fear or ignorance (Hume), or from the necessity to be completely human (Schleiermacher), or from an illusion of the world or a desire for a better one (Freud)?

Regardless of the religion we follow, no religion can ultimately replace metaphysics (the study of being), since being is more comprehensive than religion (an uncritical explanation of the cause of it). In the presence of religion or in the absence of it, metaphysics can stimulate our yearning to know, our desire to penetrate the powers of nature and the wonder of life itself. It is with the present, not the future, that metaphysics is relevant. Why would anyone abandon a certainty for an uncertainty, and why are we preoccupied with the future when the present is all that consumes us? It seems apparent that the idea of life after death came from a feeling of uncertainty about a time none of us can know, a feeling that leads us directly to the hypocrisy of Pascal, who wanted every assurance, while risking nothing.[5] Although religion attempts to answer the unanswerable, its method is not based on critical thinking.

HOW RELIGION WAS PERCEIVED IN THE PAST

When aided by archaeology, we find that religion formed an essential part of past societies. Therefore, prehistoric times seem to have differed little from historic times as far as religion is concerned. It has been said that the first tools may have embodied a religious aura,[6] but we must question this view. In what sense do tools transcend the world? Was there a mythico-religious aura when stone tools were created, when tools of bone and wood were invented, when clothing was first made, when different types of domestic implements were devised? Although these technologies greatly helped human survival and fostered well-being, none of them seem to possess an aura of religious significance. They seem not to contain any supernatural phenomenon on which religion might be based. More than these technologies, the appearance of magic and its use of natural phenomena eventually acquired a technique, and the magician, medicine man, or priest became a technician.

Because religion forms an essential part of society, we should expect to find its influence no matter where we look. We even find religion in those societies in which science was proposed as the logical explanation for the workings of the world, a notable example of which were the ancient Greeks, beginning with the pre-Socratic philosophers. Although these philosophers spoke of physical phenomena, notably, earth, air, fire, or water, as the basis of reality, they ignored any alleged religious connection between these phenomena and their origin. They assumed that the world could be explained physically, although their conception of science did not displace religion. Despite science, the Greek gods continued to be revered in ancient Greek society, and remained a source of devotion for their people. Although the Romans followed the Greeks, Greco-Roman civilization has directly and pointedly given rise to Western civilization, from which Christianization and technologization of the world (with its unique characteristics) originate.

In spite of conflicting interpretations of the fall of Greco-Roman civilization, it can be concluded that Western civilization on which it is built underwent a process of Christianization. In the later Roman Empire, from the third to the fifth centuries, the upper classes that were largely pagan were not able to withstand the influence of the more numerous urban lower classes that gave their support to Christianity. Eventually, the upper classes fell victim to the lower classes' Christian conception of the world. Although the latter absorbed the former, more importantly, Christianity also weakened the Roman Empire because individuals with a capacity for civic involvement shied away from public service. The gradual disintegration of the Roman Empire, at least in its western half, was accelerated by the decline in agriculture and the loss of its tax base. It was replaced by the intrusion of the Germanic peoples and their concept of kingship. Rome's policy of a centralized government with an efficient bureaucracy was replaced by several small, decentralized Germanic kingdoms, led by a king and supported by a warring aristocracy that became wealthy landowners. Because the Catholic Church adopted the administration of the Roman Empire at a time when the latter was in decline, and because the church assumed the leadership in the West after the fall of Rome, the Catholic Church eventually became the strongest political power in Europe.

Christianity also acquired a metaphysical underpinning from Greek philosophy. It was based on Platonism that was adopted in the first century, and continued with Neoplatonism in the third. Metaphysics that began with the Greeks as an endeavor to understand the most basic qualities of reality had an extraordinary influence on the development of Christianity. Whether or not we accept the view that Christianity became Hellenized, what resulted was the transformation of Christianity from an obscure Jewish sect into a world religion supported by a metaphysics that revealed a detailed history and a unique conception of the world. This is to say that Christianity acquired a theology with a sophisticated ontology, neither of which was included in the original teachings of Jesus Christ. Christianity also acquired a form of metaphysics from Greek philosophy that allowed it to advance a unique worldview. This Christian metaphysics became the framework to which everything else in Christian culture was appended, including science.

Although Christianity became the predominant religion of Western civilization, its metaphysics began to weaken at the end of the Middle Ages. Medieval science that became subordinate to metaphysics was challenged during the Renaissance when the importance of the individual and the rise of the middle class, both newly emphasized, began in earnest.[7] Because classical metaphysics, especially Platonism, perceived the real world to be beyond this world that we know through sense experience, and because Christianity adopted these views, a metaphysical denial of either of them began to doubt the basis of metaphysics itself and started to unravel it. Christian metaphysics conceived of a supersensory world to be the real world that was eternal, absolute, and transcendent, and the sensory world (the world we live in here on earth) became the unreal world. The basis of Christian truth was founded on alleged "facts" that were external (and therefore, untouchable historically), that is, these "facts" were so objective as to be beyond the reach of human influence.[8]

During the Middle Ages, the Christian conception of the world gave to the earth the prominent place in the universe that was supported by references to the Bible, Aristotle, and Ptolemy. This interpretation considered the earth to be in the center of all other celestial bodies. The earth was presumed to be stationary, around which the planets and the stars revolved in a finite universe, but the new science that developed at the end of the Middle Ages questioned this Christian cosmology. The truth of astronomy associated with the Christian conception of reality could no longer be upheld by the challenges brought to bear by the scientific observations of astronomers such as Copernicus, Brahe, Kepler, and Galileo. As a result, the medieval ideas of the universe that supported Christian metaphysics were undermined. The objective world order of medieval Christianity was replaced over time, as we described above, by another worldview that is temporal, relative, and immanent.[9] The transformation of Christian metaphysics did more than just question cosmology. It also signified the end of the unquestioned authority of the Catholic Church and the papacy.

It is ironic that the religious reforms of Martin Luther that began the Protestant Reformation and that challenged the authority of the Catholic Church in the sixteenth century did not so much weaken Christianity, but reinforced it. On the one hand, the religious wars fought between Catholics and Protestants that occurred in the sixteenth and seventeenth centuries seem to be unaffected by this new science, and on the other hand, the science of Christian Europe was seriously threatened by a handful of scientists, beginning with Copernicus in the sixteenth century and culminating with Newton in the eighteenth. The new science that led to the Scientific Revolution introduced more than a different theory of the universe.[10] It also introduced, within the following centuries, a new interpretation of nature supported by Cartesian physics, the comparison of the world with a machine, the prominence of subjectivity, the concept of anthropocentrism (later challenged by Darwinism),[11] the ideals of human perfection by human effort alone (especially through biotechnology), and ultimately, the dawn of nihilism.

In addition to science, technology also became a beneficiary of this new interpretation of nature. Just as changes in technology are related to those in science, technology finds expression not only because of its religious characteristics, but also because of its metaphysics. Technology acquired a metaphysical underpinning partially because of its history, and partially because it superseded Christian metaphysics. If we were to describe these ideas in another way, we should say that technology benefited from its own revolution and profited from the culmination of the changes associated with the Scientific Revolution.[12] It was only a matter of time before technology and the technological revolution would acquire predominance.

When we say that technology superseded Christian metaphysics, we mean that the dissolution of the latter led to the dawn of nihilism. Because nihilism replaced the objective worldview of medieval Christianity with a subjective worldview of a non-religious worldliness, technology became invigorated as a result. Technology does not deal with the afterlife, but with the reality of everyday existence. When the European Middle Ages began to wane, a slow and continuous process intensified both technology and science, and the latter's critical observation of the world, but the advancement of science and technology progressed at the expense of Christianity. The decline of Christian metaphysics, as discussed above, reduced the role of religion in society that has caused the weakening of religion in general. The result is society's secularization. And it is secularization that has helped to augment technology's religious characteristics.

HOW RELIGION IS PERCEIVED IN THE PRESENT

In the advancement of science from Copernicus to today, the metaphysical underpinning of religion in general and Christianity in particular have slowly weakened. Not only has science changed, but also has our worldview. An understanding of the laws of physics that began with mechanics has questioned the science of the past. In conjunction with physics, these laws also questioned its religious underpinning, since science to some extent expresses an opinion of or is associated with the religion of its time. Although religion is pursued at the present time as it was in the past, there is an essential difference between the present and the past. Because religious values are directly related to their respective societies, it can be shown that our present religious values have been rendered superfluous at worst (regardless of how much we intend to reinstate them), or privately expressed at best. Even when reinforced, these values are artificially contrived, that is, the old religions remain, but their values have been diluted.

This change in religious values has generated a secular conception of the world that has led to the privatization of religious beliefs for those that remain. Consequently, religion is rarely associated with formal religious institutions. It has now become the responsibility of individuals to be personally involved in a religion that formerly was a public ritual in a religiously conscious society. In our secularized age, people are free to follow all, some or no features of a religion, to be believers, agnostics or atheists, to believe in some form of spirituality as a life force or to deny it, to be driven by the ultimate questions of being or to live an unexamined life, to wonder about the cause of creation or not to wonder at all.[13]

Of all the religions, Christianity possesses an inherent uniqueness, and for this reason, religions other than Christianity lack the dynamics that helped to make Christianity a vehicle for technological change. The historical influence of Christianity on technology maintains not only a difference of degree, but also a difference of kind, that is, as a world religion, Christianity has achieved more than the Christianization of the world. It has, more importantly, prepared the world for technologization that we see unfolding before us. Although there are no societies that do not possess some form of technology, Christianity has become a catalyst for modern technology and its impact on the world. Hence, technological development in Western civilization is not accidental, but is the direct result, among other factors, of Christianity. It is also influenced by the Western attitude of innovation and inventiveness, as well as its positive attitude toward manual labor.[14]

Technological development is also based on and is the beneficiary of various interrelated Judeo-Christian ideas: 1) the earth is not eternal, but created,[15] 2) humans are created in "the image of God," who have a unique relationship with God, 3) they are expected to help God achieve his ultimate goal—the salvation of human souls and the attainment of the heavenly kingdom, and 4) these events will occur only once because time is perceived not to be cyclical, but linear.[16] When applied to technology, these ideas signify that the technological exploitation of the world is humanity's intended means, the betterment of humanity as the result of technology is its intended end, and both the means and the end will occur through the linearity of time, bringing to a closure alternate possibilities. This Judeo-Christian tradition helped to form modern technology. As we said above, modern technology, led by Western civilization, is no accident, but is due, in part, to the direct and immediate result of Christianity.

Religions, to some extent, are technologized because they use technology to help them convey their message. If we were to make a comparison, we might say that Christianity has not been technologized as much as technology has been Christianized. We might also add that technology's Christianization, which brings westernization in its wake, affects other religions to a lesser extent. Because Christianity influenced technology, it may be said that technology has become more Christian. Rather than Christianity surviving in a technological setting, it is more accurate to say that technology survives in a Christian setting. But what is this Christian setting like? Technology is Christian superficially, not genuinely. Although influenced by Christianity, technology will surpass it. Christianity itself will be adulterated by its technological juxtaposition so that if it survives at all, it will become fundamentalist, which as we said above, will become resistant and defiant, very much like Christianity's first few centuries in the Roman Empire in which it survived, despite its opposition, because it defied the world.

And because Christianity chooses to be different, it can reject the process of secularization, that is, the process of worldliness,[17] associated with modernity. Therein is an irony because Christian culture is the cause of its own secularization that was intensified by the Scientific Revolution and the dissolution of Christian metaphysics. Hence, the technological age is secular because its Christian roots promote it, just as Christian culture is the cause of nihilism—the interpretation that Christian morality is relative and has no absolute exclusivity. Secularization that began in the West is now evident everywhere. Whether or not it is accepted by non-Western cultures is not the issue. Whether or not it is overt is irrelevant. By a process of osmosis, secularization finds a way to invade everything. Its very presence throughout the world challenges all existing cultures, and therefore, all existing religions.

But there are other ideas that must also be discussed. In addition to Christianity being the catalyst for technological change, technology itself has characteristics that enhance its own religiosity. If religion develops by accumulating "a body of revealed truth[s],"[18] then there are similarities between this idea and the idea that technology itself possesses characteristics ("truths") that will transform it into a religion. These characteristics are numerous and have already been discussed in this book. But enumerating them again would be helpful. They include the following: technology's techniques have become its rites of passage, its doctrine is preserved in scientific and technological rationality, its catechisms or books of instruction are printed as technical journals and specialized manuals, its temples are manifested in the academies of science and university laboratories, its ecclesiastical councils are represented in professional conferences, its priesthood is evident in the expanding body of scientists, technicians and engineers, its sacred history is revealed in the idea of progress, and its theology is its accumulative presence in the world. At some point, it may develop a holy book that will contain its sacred texts.

Like Christianity, technology also proselytizes the believers of other religions. Although these characteristics have existed for quite some time, technology lacks a prophet who will speak its truth and reveal its divinity. Just as the Old Testament predicts the coming of the messiah, technology can do the same. Each day brings us closer to this realization. It is demonstrative of our worldview that has made technology the predominant phenomenon of our age. Its religious form is dedicated *ad majorem technologiae gloriam*.

Nor is technology removed from human experiences. In fact, it influences human experiences on the grand scale. And therefore, technology is a worthy candidate for a religion because it is not removed from those experiences that help to develop it. It is tied to the world by means of its praxis, reinforced by its practical application in everything we do. Because it is ubiquitous, technology lends itself to a presence that we create. It is a simple task to pass from its ubiquity to its divinity, but it contains an inherent contradiction that concerns the following idea: since technology deals with the reality of beings and since God or gods are non-beings (because they are not things), how can technology become a new religion? Either we must admit that technology is an unworthy candidate to be a true religion, or we must accept technology as the means in which we are tied to the world as the only realm of being. This realization would require a different worldview from the worldview that conventional religions confer.

More than a valid attempt to reveal reality, technology will introduce a different kind of religion—a religion that is non-transcendent and based solely on sense experience, leading to the creation of empirical deities. As a religion, technology does not conceive of a world different from the world we live in. The supersensory world of Christianity or Islam or Buddhism will be challenged by the sensory world of technology. Contrary to the idea of religion in general, technology will produce no transcendental deities because it is tied completely to the world. This understanding is augmented by nihilism, which as we said above, replaced an objective and absolute worldview with a subjective and relative one.

HOW RELIGION MIGHT BE PERCEIVED IN THE FUTURE

Irrespective of the threat of fundamentalism, existing religions will give way to the new. Old religions still worshipped today that first made their appearance in antiquity may be followed by a new religion that could appear everywhere, a religion that will most likely make its appearance first in the West because that is where Christianity is most developed and where advanced technology is found. These old religions were created in former ages that are now being surpassed in an age given, in addition to the concepts described above, to homogeneity, mechanization, and human domination through biotechnology and genetics. The fact that we can predict the coming of a new religion does not mean that everyone will accept it unconditionally, just as everyone did not readily accept the old religions. Because the old religions are so well entrenched may be reason enough to assume their perpetuation, especially in remote areas (geographically or culturally) of the world. Nevertheless, the challenges of modernity will gnaw away at the edges of these religions until what remains will be an empty shell. All gods die, and some religions evolve into others.

The old religions made its believers feel powerful, protected their cities, made their women fertile, and explained what lies beyond death. A new religion may also do some of these things, but it will also alter the effects of aging, cure disease, promote physical and psychological well-being, and triumph over life's infinite flaws. Although we may be in awe of the power of a new religion, can it help us achieve great things? Its care and concern may help us prevail over obstacles, but can it foster a worthwhile morality? Nevertheless, creating a new religion is not unreasonable, since it has occurred many times in the past. Because no religion is completely acceptable to everyone, believers of one religion disavow or oppose those of another. There are religions that perceive their god(s) to be objective and external, as there are religions that have no god and base their beliefs unrelated to any transcendent. Because religion conceives a world beyond reality, it is often confused with metaphysics, although one task of the latter should be to remove misconceptions that have accumulated over the ages by religious distortions and philosophical impositions. In this sense, metaphysics can survive in a secular age.

The concept of God or gods is also associated with the way of life that has made the world we know. Technology as a new religion is opposed to the idea of a personal god that is fundamental to Judaism, Christianity, and Islam. A personal god may be comforting to its believers, but it challenges the idea that the world, as a particular state of being, may have originated as the result of the general phenomena of life, or the effect of natural processes that are governed by their own laws.[19]

Although technology promotes anthropocentrism, this idea does not mean that technology is capable of creating ourselves artificially, as robots are made in the laboratory. Nor does it mean that technology's religious form should be associated with spiritualism, millenarianism (or Chiliasm), Rosicrucianism, science fiction, UFO's, the Raelian movement, artificial intelligence, scientology (L. Ron Hubbard), dianology (Jack Horner), abilitism (H. Charles Berner), amprinistics (Harold Thompson), or Teilhardian phenomenology. Nor should technology be confused with secularism, that is, the worship of the secular (when secularism is considered religiously), or the emphasis of the worldly (when secularism is considered non-religiously), although secularism is related to worldliness that accompanies technological development.[20] Given technology's inherent characteristics and the influence of Christianity, it is evident that technology has the potential to develop into a religion.[21]

When referring to technology as the vehicle by which a new religion will come forth, we cannot ignore the scientific evidence that seems to indicate that life on the earth, including humanity, evolved from earlier life forms. The fossil record points to evolution as the likely explanation for all life on earth. Technology was created by humanity to assist in human survival, that is, humanity appeared first and technology followed. Technology did not create us; it merely supplements what life has to offer. It appeared long before the first religions were conceived. Because technology resides behind, beneath, above, and beyond anything we experience in nature, we might say that it has supernatural-like characteristics.

By giving answers to basic questions about life and death, religion confers a sense of security, enabling us to live our lives with a feeling of stability. For those religions that have survived into our time, their meaning has been diluted. Fundamentalism is one way in which this tendency is resisted. Although there are individuals who tenaciously hold on to their religious beliefs, they are faced with few options, one of which is the denial that their religious values have waned. Another alternative is "to be born again" that results when those who formerly abandoned their religion use it to seek self-identity. Because there is a tendency for some people to react violently against technology, some religions may survive in a sub-culture. When this situation occurs, there are similarities between technology's future religion, accompanied by archaic religious remnants of the past, and the Christian Middle Ages, accompanied by pagan survivals, such as paintings, objects, rituals, or festivals from pre-Christian times. Regardless of the form technology takes, its theology can include this dictum: no salvation (that is, no benefit) outside of technology, which is similar to Christianity's old dictum: no salvation outside of the church. Technology can become the self-appointed means to salvation.

It also seems likely that future technological societies may create a need for many gods. In this sense, the new religion would be similar to Hinduism that has no one principal god and yet hundreds of subordinate gods. We should also remember that monotheism is not necessarily an advancement over polytheism because there have been societies in which polytheism followed monotheism, and still other societies in which the reverse occurred. We make this conclusion because most conventional religions are monotheistic. But polytheism is a more natural form of religion, in that each force, like the wind or the rain, is considered to have its own creator, although a single god could emerge predominant rendering all other gods subordinate. Hence, a cyclical phenomenon of religion emerges: polytheism could give rise to rivalry among the gods, from which monotheism could originate. And monotheism could dissolve into polytheism, which then completes the cycle.

We should wonder how existing religions with the fear of damnation, as preached by Christianity or Islam, could be replaced by a technological religion that may seem to some people to be largely a human fabrication. Although some might say that this new religion is irrational, if not blatantly blasphemous, others might say that it is another example of human creativity, and indicates once again that religion is not found, but created. Religions are similar in so far that they attempt to offer an explanation of the world and our place in it. Although religion in general may contain inherent contradictions, there is a need for it because it is an attempt to understand reality. Most people attempt this understanding through a religious context, as we said above. If we recall Schopenhauer's definition of religion as the way in which people understand being, then we should be able to understand the meaning of this statement. Hence, technology can legitimately and willfully be offered as a new religion. But the problem remains that if we cannot create a religion that advocates a worthwhile morality with a set of beliefs that establishes meaningful social institutions, then the only alternative is the pursuit of a lesser imperative that may be more accurately described as mythology. Religiously, we live in a formative time when the old religions are eroding and the new will take their place, a time when the faithful will give reverence to the new gods, and will ask for their blessings.

CONCLUSION

Just as Christianity became a religion and acquired a metaphysics, technology can do the same. When discussing technology, we have to be mindful both of its history (because it is the result of an accumulative process), and its metaphysics (because it displays an intricate presence in the world), but religiously, technology is a threat to other religions. Because of its history, technology has succumbed to the cultural influences and progressive ideas of Christianity as it evolved through the Middle Ages. Christianity has helped to make technology a potent force with a dynamic energy before which everything submits. The Western spirit of innovation, the importance of manual labor, and technology's own inherent characteristics, all of which intensify technology's religiosity, have also supplemented Christianity's importance.

Although the discussion above concerns the importance of metaphysics in a technological age, its survival is dependent on the removal of idealistic or anti-empirical characteristics that it acquired over the ages. These characteristics have distorted metaphysics and have made it a near equivalent to theology. It has been said, particularly by Heidegger, that we have reached the end of metaphysics because, following Nietzsche's interpretation of the "death of God," the dissolution of Christian metaphysics means that metaphysics in general has been undermined.[22] We need to question the meaning of the end or completion of metaphysics. Rather than representing a termination, we should understand this expression as a transformation, a radical change from the interpretation of reality put forth by Plato's idealism, and later discredited by modern science. We must reiterate that metaphysics has validity, not only because ontological freedom is the basis of human reality, but also because the most profound manner in which we may understand reality is by asking fundamental questions about the nature of being. What has changed is not being, but its interpretation.

If the end or completion of metaphysics means the rebirth of religion because it is now possible in our secular age to believe again, this condition would be preparatory to the revival of existing religions and to fundamentalism, the first of which is proactive to modernity and the second of which is reactive. But the rebirth of religion would also be beneficial to technology, which now has a chance to be self-fulfilling.

More importantly, the dissolution of Christian metaphysics has enabled technology to become a beneficiary of nihilism that concerns the negation of traditional values, or at least the sublimation of higher values to lower ones. Nihilism has appeared because the dissolution of Christian metaphysics has left a void that is now occupied by technology, that is, the rise in the importance of technology has been caused by the decline in the importance of Christian metaphysics that means, of course, the decline in the importance of Christianity. Within this void, technology may evolve into a religion. But we should emphasize that religions need to enforce meaningful moralities whose social applications benefit society without falsifying a fair and reasonable conception of the world. How well technology addresses these ideas will reveal how meaningful it will be as a religion.

NOTES

1. George Santayana, *The Life of Reason, or the Phases of Human Progress*, 5 vols., New York: Charles Scribner's Sons, 1905-1906, III (1905): *Reason in Religion*, pp. 9-10.

2. William James, *The Varieties of Religious Experience: A Study of Human Nature*, New York: Longmans, Green & Co., 1902, pp. 250-253 and 478-484. Also see David Hume, *An Enquiry Concerning Human Understanding*, sect. XI, marginal sect. 105-106 in *Enquiries Concerning Human Understanding and Concerning the Principles of Morals*, ed. L.A. Selby-Bigge, 3rd rev. ed. P.H. Nidditch, Oxford: Clarendon Press, 1975, pp. 135-138.

3. Since the English language uses the same word for individual beings as well as the totality of them, the use of a capital "B" for the latter may confuse this term with "God."

4. Arthur Schopenhauer, *Parerga and Paralipomena: Short Philosophical Essays*, trans. E.F.J. Payne, 2 vols., Oxford: Clarendon Press, 1974, II, p. 325.

5. See sect. 418-233 in Blaise Pascal, *Pensées* in *Oeuvres complétes*, préface d'Henri Gouhier, présentation et notes de Louis Lafuma, Paris: Éditions du Seuil, 1963, pp. 550-551. Also see *Pensées*, trans. A.J. Krailsheimer, Harmondsworth: Penguin, 1966, pp. 149-153, where sect. 418-233 is described as "the wager."

6. Mircea Eliade, *A History of Religious Ideas*, trans. Willard R. Trask, *et al.*, 3 vols., Chicago: University of Chicago Press, 1978-1985, I (1978), p. 6.

7. This fact does not mean that the importance of individuality originated with the Renaissance. For the concept of the innate value of each person and its relationship to individuality, see Colin Morris, *The Discovery of the Individual, 1050-1200*, London: Society for Promoting Christian Knowledge, 1972, rpt. Toronto: University of Toronto Press, 1987, pp. 65-86.

8. Therefore, objectivity, as one characteristic of Christianity, described an eternal realm of being that was unchanging, and within this realm were to be found heaven and hell. The interpretation of this supersensory world at the end of the Middle Ages underwent a process of transformation, resulting in its opposite: the triumph of subjectivity. When the world began to dissolve into subjectivity, that is, into a subjective interpretation of reality, what are we describing? Should we use the term "interpretation," since reality no longer has a measure acceptable to everyone, or should we use "technique" that can more accurately describe the same reality? It is through technique that we render an interpretation. So it is not hermeneutics (the science of

interpretation), made famous by Hans-Georg Gadamer, that we need to subscribe to, but the application of technique that lies at the foundation of technological society that has become the new vehicle to truth. Truth is now defined by a new paradigm—the paradigm of technology, to which science is appended. Whether or not technology is the best criterion to assess truth is irrelevant, an idea that also pertains to the relevancy of the previous criterion—Christianity. Historically and metaphysically, the first paradigm gave way to the second.

9. For an excellent description of Christian metaphysics, see Wilhelm Dilthey, *Introduction to the Human Sciences*, ed. with an Introduction [and Translation] by Rudolf A. Makkreel and Frithjof Rodi, Princeton: Princeton University Press, 1989, pp. 174-197.

10. This new theory of the universe is discussed in Thomas S. Kuhn, *The Copernican Revolution: Planetary Astronomy in the Development of Western Thought*, Cambridge [Mass.]: Harvard University Press, 1957, pp. 229-265.

11. Although geocentrism (the placing of the earth at the center of the universe) is associated with anthropocentrism, the destruction of geocentrism that began in the sixteenth century did not remove anthropocentrism as a principal characteristic of modern technology. In fact, anthropocentrism was reinforced by Cartesianism (which appeared in the seventeenth century) because it emphasized the importance of subjectivity as an abiding characteristic of the modern age.

12. Hans Jonas, "Seventeenth Century and After: The Meaning of the Scientific and Technological Revolution," in his *Philosophical Essays: From Ancient Creed to Technological Man*, Englewood Cliffs: Prentice Hall, 1974, p. 48.

13. In general, see Gianni Vattimo, *After Christianity*, trans. Luca D'Isanto, New York: Columbia University Press, 2002.

14. Christianity attacked the institution of slavery that it inherited from antiquity, and replaced it with the importance of manual labor that Roman society disapproved.

15. To which we might add that because the earth is created, it is transitory.

16. These ideas are also discussed in chapter two above and are derived, in part, from Ernst Benz, *Evolution and Christian Hope: Man's Concept of the Future from the Early Fathers to Teilhard de Chardin*, trans. Heinz G. Frank, Garden City: Doubleday, 1966, pp. 121-132. Benz's research is also discussed and supported in Lynn White, Jr., *Medieval Religion and Technology: Collected Essays*, Berkeley: University of California Press, 1978, pp. 235-237.

17. A critic of secularization and any theory associated with it is Olivier Tschannen, *Les théories de la sécularisation*, Geneva: Librairie Droz, 1992. When considered from a non-sociological point of view, secularization is concerned more with a qualitative rather than a quantitative perspective, and for that reason, it has its supporters, notably Hermann Lübbe, David Martin, and Richard Fenn.

18. Eric J. Sharpe, *Comparative Religion: A History*, 2nd ed., London: Duckworth, 1986, p. 48.

19. See Ludwig Wittgenstein, *Tractatus Logico-Philosophicus*, trans. C.K. Ogden, with an Introduction by Bertrand Russell, London: Routledge & Kegan Paul, 1922, 6.423-6.522 that describes many spiritual and mystical ideas.

20. For the differences between secularization and worldliness, see Hans Blumenberg, *The Legitimacy of the Modern Age*, 2nd rev. ed., trans. Robert M. Wallace, Cambridge [Mass.]: MIT Press, 1983, pp. 3-5, 8-10, 37-51, 74-75 and 116-120.

21. Since it is inferred by some sociologists that secularization causes a religion's resacralization, should we ask if secularization is subject to cyclical rhythms? This cyclicity is upheld by Rodney Stark and William Sims Bainbridge, *The Future of Religion: Secularization, Revival and Cult Formation*, Berkeley: University of California Press, 1985, pp. 2-3.

22. This idea is derived from Friedrich Nietzsche, *Gesammelte Werke*, Musarionausgabe, 23 vols., Munich: Musarion Verlag, 1922-1929, XII (*Die Fröhliche Wissenschaft*), no. 125 ("Der tolle Mensch"), and rendered into English as "the madman" in *The Gay Science*, trans. with Commentary by Walter Kaufmann, New York: Vintage Books, 1974, pp. 181-182. Also see Martin Heidegger, "Nietzsches Wort: 'Gott ist tot'," in his *Holzwege*, Frankfurt: Klostermann, 1950, pp. 193-247, and rendered into English as "Nietzsche's Word: 'God is Dead'," in *Off the Beaten Track*, ed. and trans. Julian Young and Kenneth Haynes, Cambridge: Cambridge University Press, 2002, pp. 157-199. Also see the criticism of this essay in Karl Löwith, *Martin Heidegger and European Nihilism*, ed. Richard Wolin, and trans. Gary Steiner, New York: Columbia University Press, 1995, pp. 96-127.

Chapter Five

Technological Theology

INTRODUCTION

Although not deliberately combative, technology has the capability of developing a theology that is derived from its accumulative presence in the world. Whether from existing religions or from technology, a theology is the effect, not the cause, of its presence, but this effect is especially true for technology. Because any circumstance or situation of a religion is not a given, a religion is often related to conditions beyond its control. These circumstances or situations are also tied to the revelations that allegedly legitimize a religion and supposedly strengthen the belief that follows from it. For example, Christians accept the changing of water into wine by Jesus during the marriage feast at Cana as proof of his divinity. This and similar revelations contributed to the religious aura that evolved around the person of Jesus Christ.

Religions seek revelations, but the present age is unique because even technology is characterized by revelations that lay a basis for its religiosity. Revelations and their religious importance should not be a source of confusion. Either revelations exist and people believe in them giving a footing to a religion, or they exist and people ignore them, or they do not exist. To these possibilities, technology is a notable exception because the existence of technology cannot be denied by believers and non-believers alike. Muslims and atheists, Jews and agnostics, Christians and scientologists support it. Because it is evident from the equator to the poles, technology's applications are ubiquitous, and therefore, it is only a matter of time before its revelations may be accepted as proof of its religiosity.

Technology has developed over vast expanses of time that began with the dawn of humanity. It is supplemental to human evolution and to the latter's domination of life on earth. But technology has also been accelerated by events that occurred during the Middle Ages that predate the effects of the Scientific Revolution of the sixteenth century. These facts are evident despite the understanding that modernity in general has deep medieval roots. How these roots give sustenance characterize the modern age. Since the sixteenth century, technology has also benefited from the decline of Christian metaphysics and the metaphysical void that followed, into which nihilism has flowed. These historical and cultural factors cannot help but to intensify the importance of technology.

Because technology is predicated on the realization of its future activities, it is a teleological endeavor. It achieves this realization because it projects its results into the future, but its teleology is also augmented by Christianity, which as a progressive religion, has had a lasting influence on technology since antiquity. Christianity's one-dimensional progressiveness has made technology a dynamic force in the world. These ideas reveal that technology's teleology is both internally defined and externally directed. It is internally defined because its influence is implied in its being, since it is of the nature of technology to alter the shape of things to come. It is externally directed because of the influence of Christianity that has had a strong impact on the development of technology since Christianity's inception. Because technology's teleology is culturally and historically instructive, the manner in which technology reveals itself is indicative of its being. Therefore, it is an understatement to say that technology's teleology is metaphysically pertinent.

It is also an understatement to say that although technology's theology is influenced by its metaphysics, its metaphysics is not its theology. Nor is the metaphysics of any religion equal to its theology. Metaphysics concerns the most fundamental characteristics of being that reveals its identity, discloses its manner of existence, and manifests its presence. Unfortunately, the theology of any religion is also presumed to reveal, disclose, and manifest. But there is a substantial difference between metaphysics and theology because the latter is supplemental to the former. Certainly, a divine presence must exist first before it chooses to disclose itself. A miracle would be a common example in that something extraordinary is considered as proof that not only does a supernatural being exist, that is, it has a presence, but this being also wishes to make this fact known to us, that is, it wishes to communicate with us. This self-revelation coincides with a basic understanding of theology, and attributes speech and deeds as revelatory events.

Anything that exists can have a metaphysics, but metaphysics and theology are not coterminous. Although theology might not interest many people, religion does. In fact, religion has become the means by which many people interpret reality.[1] Any priest or rabbi, minister or imam, monk or nun, may declare that the world was created by divine will, but this declaration does not affirm reality. Supposedly, we are past the time when any unfounded or unproven presuppositions should be accepted without question, although religious believers may still follow certain declarations, despite scientific reasoning, or methods, or discoveries that indicate otherwise.

And yet theologians are the first people to admit that theology is a human creation. They should also admit that organized religions have been invented as long ago as the Neolithic age. Before organized religions, people worshiped the spiritual world with gods that represented the embodiment of natural forces, or cosmic energy. And when gods are made in our own image, of which there are many, this anthropomorphism intensified the idea of our own self-importance.[2] With them, we look to a power that originated in ourselves, but that we now impart to the gods. Gods express this capability as if they were fulfilling our greatest desires and disclosing our greatest attributes. A warlike people, for example, would have little need for a god of peace, since they emulate the attributes of war. Likewise, a people who think they have been especially chosen by their god would oppose the idea that their god is not special for choosing them. Whatever suffering these people endure in the name of their god is accepted as proof that indeed they are chosen.

Religions since the Neolithic age deal with the idea of human imperfection that presupposes the need for improvement. These religions were influenced by the development of self-consciousness that is thought to have originated in the first millennium before Christ, a time described by Jaspers as the "axial period."[3] This period is perceived as a pivotal time in which humans became aware of the relationship between themselves and the world around them, and concerned new ideas regarding humanity's ontological limitations. The solution that was proposed to overcome these limitations involved the invention of gods, that is, the latter were powerful external forces that embody qualities we emulate. Organized religions, even those that have appeared centuries later after the beginning of the axial period, such as Islam, have achieved success because they have subordinated individuals to gods, that is, to abstractions of the highest order. Although it is difficult to identify the precise beginning of the axial age, it evolved slowly over several centuries. These religions developed evaluative judgments of human imperfections that were derived from an inherent vulnerability.

But technology is different from these religions. Of all the characteristics of modern life, it is technology that has attained the most important influence. Because of the factors discussed above, it is likely that technology will supersede organized religions and will replace them with itself.[4] Like organized religions, technology has the potential to subjugate individuals, but unlike organized religions and the religious premises on which they are based, technology is no abstraction. Because technology is tied to reality, it is unlikely that a golden idol will be erected for its adoration.

The possibility that technology may become a religion may occur simultaneously with the fact that religion in general is considered by many people to offer an explanation of the nature of being, that is, religion has been regarded to be the equivalent of metaphysics. There are even those who regard the predominance of technology to be the final phase of metaphysics because the being of human beings is sublimated to technology.[5] Despite the notion that religion renders a metaphysical interpretation of the world, there is some truth to the idea—at least in regard to technology—that indeed the latter represents reality, particularly the way the present age perceives it. And therefore, technological theology supplements technology's metaphysics.

Each religion introduces a different interpretation of reality, and yet, religions are similar in so far that they are associated with and supported by a community of like-minded believers, adhering to specific beliefs, and following distinct ideas. The religions of early civilizations and the modern world offer interpretations that are similar, but have obvious differences. Since there have been many religions, it would be pointless to ask if religions are the result of human imagination. Instead, we should ask what they reveal about the human condition. Can we conclude that they reflect specific desires and fears as determined by a culture's metaphysical needs? For example, was the religion of nomadic Paleolithic hunter-gatherers more meaningful to their culture when compared with the religions of settled agricultural people? Was the religion of pre-Christianized Vikings more beneficial for their way of life when compared with the religion of the Christians they attacked?

Although cultures change, the characteristics that are reflected in the formation of any religion are slow to change. Religions reveal the past. They are time capsules, presenting a portrait of a people, like a photograph, exposed by the light of day that it now buried by time. They also reveal the customs of former peoples. It is not true that religions, in general, have adapted their customs to the world, but that the world is expected to adapt its customs to religions. This is to say that any age is expected to conform to its past, and if an age does not conform, it suffers. The exception is technology because it conforms to every culture and is expressive of any age. It may become the universal religion. The permeation of every aspect of reality, that is, the accumulative presence of technology, makes it easier for technology to be accepted metaphysically and religiously. It is a reality that few can deny, but a reality that is innately both a blessing and a curse.

Since the first criterion of any religious belief is trust, we either trust that witnesses and martyrs, or saints and confessors are verifying truth, or they are not. Religion is dependent on its sources, but it is also dependent on the acceptance of them. Since a follower's belief in any religion excludes belief in another, it is belief that makes or breaks a religion. A religion ceases to have an influence when there are no longer any believers, either because they no longer promote its truth or they have died. This phenomenon is true for all religions and also applies to the religious truth of technology. It is believers who give a religion its identity. Belief may be a certainty for those who have faith, but its proof remains uncertain, or to use the language of Kant: Belief is subjectively certain, but objectively uncertain. Although revelations may be explained by parapsychology as phenomena that originate from abnormal behavior, the importance of revelations is still affirmed by those who believe in them.

THE NATURE OF THEOLOGY

It is sometimes said that theology is a Christian invention, although the word itself is of ancient Greek origin. And since the time of the Greeks, all major religions have developed a theology, to which may be added liberation, creation, quantum, or feminist theologies. Therefore, the nature of theology has changed, and it has changed primarily because of its widening influence.[6] Etymologically, theology is derived from the Greek words *theos* (god) and *logos* (word), meaning the study of the nature and attributes of God or gods.

The ancient Greeks discussed among themselves how theology might be studied. Plato said the attributes or the nature of the gods could be studied, which signified that theology is a practical endeavor.[7] Aristotle said theology is a theoretical science, that is, it begins with the subject matter (something that is known) and not with the person who wishes to know.[8] This difference in interpretation seems to reveal more about the opposing views of these two philosophers rather than the nature of theology, since Aristotle perceived the phenomena of reality to be tied to the world (phenomena that are objective and empirical), but Plato perceived that a real world exists, except that it lies beyond this one (phenomena that are idealistic and supersensory). Regardless of this difference of opinion, theology has become a reputable endeavor for many religions. And it is largely an academic endeavor.

Theology is upwardly mobile because, as an inductive discipline, it starts in the world that supposedly leads to the divine. It is derived not just from conclusions based on individual revelations or events that are thought to be divinely inspired, as the multiplication of oil for the celebration of Chanukah, but also from personal reactions to them. The events of theology are derived from a variety of sources: voices from above that are interpreted from below, voices from within, dreams, and visions. This is to say that revelations are private matters that make them difficult, although not impossible, to scrutinize. Revelations do not concern universal experience, but particular descriptions of events. The human condition that unites all people contains an inherent universality that is not influenced by the particularity of revelations. We mislead ourselves if we think that particularities have any bearing on universality. Theology is also devised by individuals who are engaged with these revelations and with fellow believers in a culture and at a time when a description of these events portrays an underlying need for them. And need justifies acceptance. In fact, what is a need if not an openness to being—physiological, emotional, social, ontological.

Traditionally, theology also concerns the interpretation of religious texts that contain descriptions of divine revelations. It is the prerogative of a prophet to write texts that become the focus around which a religion revolves, such as the Vedas or Upanishads to Hinduism, the Old Testament to Judaism, or the Quran to Islam. How these texts are interpreted reflect their relevance to changing times. Therefore, hermeneutics or the science of interpretation is a part of theology. It is an evolutionary science. It is based on the technique of guided reading and is influenced by critical analysis. Scientific explanations of the circumstances that may be interpreted as revelations, however, do not and cannot negate the need of some people to continue to believe in them.

But more than hermeneutics, new ideologies have also transformed the traditional understanding of theology. This explains why theology has expanded into areas hitherto unknown. When we discuss theology, we cannot assume that it pertains exclusively to Christianity, or Judaism, or Islam. Although theology still concerns the nature of God or gods, its scope has expanded to feminism and the innate worth of the humanity of women (feminist theology), religious interpretations of the science of quantum theory (quantum theology), attempts to serve the poor and to relieve the effects of poverty (liberation theology), respect for the earth (creation theology), and, of course, technology (technological theology).

Theology has devised arguments for and against a religion's realization that brings to mind the difference of opinion between Plato and Aristotle discussed above. Is knowledge of God or gods part of all knowledge? Is reason, which is a universal human trait, related to theology, even for those who deny the latter's importance? Is belief, which is an essential requirement of religion, related to reason? Regardless to any school of thought that struggles with the idea of faith and reason, technology offers answers to these questions. Certainly, we can say that theology, that is, knowledge derived from the study of God or gods, is presumed to be a part of all knowledge. In what sense is truth embodied in technological theology? In what sense does technology manifest phenomena that make evident a claim for a theology?

TECHNOLOGICAL THEOLOGY

Contrary to many religions, technology as a religion will come into play not by exclusion, but by inclusion. And whatever is included is not only the obvious effects of technology, comprising inventions, products, and devices, but also technology's wide-range of methods and organizations that permeate every aspect of human reality. As a religion, we cannot describe technology in a deconstructive manner, but by the opposite, as an all-encompassing phenomenon that is an integrated part of life itself. It penetrates all of human reality. The religiosity of technology manifests itself in every place where the human mind and hand exert themselves. Because technology is open to the world, it lends itself easily to communication. And communication is a key concept of theology. What would be the point if we could not communicate with the gods? Although technology can function in a closed environment, its isolation is rare. It, too, is illustrative of communion with us.

More than the religions of the past, technology is evolving into the religion of the future. When applied without careful consideration of its consequences, technology is harmful, and when applied with careful consideration, it is helpful. Technology is not neutral, but bears the moral consequences of its applications. Similarly, a religion is not the basis of morality, but the result of the moral choices that underlie it. It is not the cause, but the effect of morality. Religion is based on moral choices that support the morality of the religion that promotes them. Therefore, the relationship between moral choices and the morality of religion is cyclical. And the use of technology conforms to this principle. How we use technology will determine if we are wise or foolish with this principle. Moral choices, when divorced from organized religious systems, may signify that the use of technology can be free from antiquated worldviews. Whether or not we will make the best choices or will avoid aggressive behavior remains to be seen.

Like any religion, technology's religiosity is presumed to be evident in the world, but unlike these religions, technology's religiosity in fact is evident in the world. To deny this truth is to deny reality.[9] How does technology manifest religiosity? How does it reveal its religious aura? Contrary to existing religions, technology's religiosity is not otherworldly, but worldly; not transcendental, but immanent; not vertical, but horizontal. Although universal, its religiosity is particularized in individual cases. Unlike religious anthropomorphism, technology is not an abstraction. Its disclosure is a self-revealing manifestation that is pointedly apparent because it embodies reality. Although trust is the first criterion of any religion, trust in technology is different because its confirmation is not subject to doubt, that is, there is no doubt that technology is in the world. We may doubt the significance of the burning bush, or the resurrection of Jesus Christ, or the guidance given to Muhammad by Gabriel the Archangel, but there is no doubt about the presence of technology and its impact. Technology as a religion is lived in the world because it is of the world. It addresses human vulnerabilities and attempts to alleviate their shortcomings within a worldly context.

Theology is not legitimized by history, and neither are revelations. But contrary to Gotthold Lessing,[10] technology's historical truths can legitimize its theological truths, and these historical truths concern its accumulative presence at all times and in all cultures. Technology has now come to mean the domination of all life on earth. This fact is true because technology has always been actively engaged with the world. It is obvious that we trust technology, and through trust, we apply it. For what is an application of anything if not a recognition of the existence of that thing? Application denotes practicality that should remind us that, along with sociability and belief systems, practicality is one of the essential characteristics of any religion.

Because technological theology is unrelated to nature, it shares a similarity with the thought of Hume who rejected natural theology as proof of divine existence.[11] Natural theology presumes that the existence of God or gods can be proven because of the existence of nature. If nature exists, then it is assumed that some divinity must exist in order to create it, and this relationship is presumed to occur even without the need for revelations. Although technology shares a similarity with Hume's rejection of natural theology, its rejection is based on criteria that concerns the very essence of technology, that is, technological theology rejects natural theology because there is nothing natural about technology. Its essence is derived from its application that is based on an artificiality not to be found in nature. Although technology relies on nature, it remains unnatural. It may even be said that humans are part of nature, although long ago they lost their naturalness.

As we said above, religions strive to have a theophany or revelations of divine events, but even before an attempt has been made, technology's revelations already demonstrate its religiosity. These revelations are evident wherever we turn, since technology's omnipresence lays the foundation for its theophany. Apart from its impact, technology is also omnipresent because its presence is tied to ours. Unlike a traditional interpretation of the meaning of divine presence, technology is subject to an empirical interpretation. Technology compels a self-examination that is far more critical than the self-examination of any religion because it leaves nothing to chance. Divine revelations, particularly those associated with Judeo-Christian tradition, are thought to be unsurpassable because they are tied to the ideas of salvation and damnation, beyond which there is no recourse. However, before these ideas can be verified, proof must be offered first that indeed there is life after death.

We have also allowed technology to be omnipotent. Its omnipotence is not only an aid to survival; it is also a feature of its omnipresence. Technology originates in this world through human ingenuity. Although it can influence what is beyond nature, it does not originate there. The presence of these revelations, which are demonstrations of technological reality, may lay the basis for religious belief. Contrary to most theologies, the theophany of technology is not tied to a realm beyond it or to supernatural beings that transcend it. Humanity is responsible for what it does. Since technology is based on reason, apart from other factors that are called into play, it would make sense that we use reason to the utmost without threatening existence.

Another requirement of theology is preaching and teaching of religious truth. Because any religion contains these two interrelated activities, it also applies to technology because its presence is everywhere, as we described above, from the instructions that accompany every product to the effects of all methods and organizations that accompany every human activity. Technology preaches and teaches more than any organized religion, but it does so unintentionally. It does not have to proselytize. As already discussed, technology is inclusive and permeates all aspects of human reality because of the nature of its being. If there is anything in the world that is not influenced by technology, it is only a matter of time before it will be.

A characteristic of theology is its intercession for believers because the latter intercede for it. Because theology is upwardly mobile, revelations must originate in the world before they can give credence to their acceptance. And after they are accepted, their importance is assured, that is, acceptance makes them important. To deny their importance, even to deny their existence, is unthinkable to believers. Even heretics do not abandon their beliefs without a struggle. When making comparisons to technology, we can say that because we believe in technology, this belief renders its religious truth believable. One can deny that technology has religious importance, but one cannot deny its revelatory importance. One can deny its divinity, but one cannot deny its presence. One may even be so bold as to deny its theology, but one cannot ignore its metaphysics.

Interceding for theology is related to the propagation of a creed, which as a statement of religious belief confesses that belief. Christianity, for example, promulgated its creed in the fourth century and helped to establish Christian orthodoxy. Although Christianity originated in the first century, its creed evolved slowly, taking three centuries to perfect. Its creed indicates Christianity's contentious nature. On the one hand, technology's creed has yet to be written, and on the other, it is questionable whether or not it will need one. Judaism and Islam have no creed. Since creeds are promulgated because of controversy, the absence of the latter renders a creed unnecessary.

One important aspect of theology concerns the human confession of wrongdoing and the divine policy of forgiveness, the latter of which is one of the greatest attributes of any religion. It concerns the mechanism of sin and how to remove its blemish. Because we often attempt to correct our mistakes, even our technological blunders, in a sense we acknowledge sinful behavior. Improvements in technology are related to the use of reason as they are also related to historical choices and cultural peculiarities. Although moral progress might be an illusion, technological progress is a reality. Like religion, it is of the nature of technology to overcome its shortcomings, that is, we should use technology wisely and seek the most helpful avenue of its application. As far as our sins are concerned, that is, those mistakes we used to confess to a priest, we now confess to a psychologist, or engineer, or doctor, who like a priest, is a technician.

Although there is no proof of life after death, religions place heaven and hell in a world to come that has similarities to the supersensory realm of Plato and the latter's conception of metaphysics. A religion without an afterlife (or a higher state of consciousness) is regarded by theologians to be a contradiction because it denies the need for salvation and therefore challenges a religion's legitimacy. But even early Christianity, and the many forms it took, formulated a doctrine in which divine inspiration was thought to be found within each believer totally unrelated to the idea of the end to the world and the immediate resurrection of all believers. Speculation about the world to come has been disguised as fact, but if an example of heaven and hell can be found in this world, either of them can be achieved through technology that can be used to improve the world (heaven) or to undermine it (hell).

CONCLUSION

Unlike the Christian concept of eschatology that concerns the fulfillment of Christ's mission or projected goal at the end of history, technology's mission is revealed within history. This revelation is additional proof, contrary to Lessing, that the historical reality of a religion, which in this case is technology, parallels its theology. But technology's accumulative presence does more than demonstrate its theology. It also validates it. Of course, like any religion, technology needs a leap of faith in order to become a religion. Once behind its protective walls, one can take refuge. These ideas give rise to two questions: When will this occur? And how will it take place?

In order to be genuine, technological theology must proceed at its own pace. Although influenced by Christianity as technology evolved in Western civilization, Christianity should not be assumed to be the basis for technological theology. Despite being historically relevant, Christianity is not relevant metaphysically. A theology may even be anti-Christian and still be genuine. Christian theology is derived from the changes that occurred in its past, and influenced by its struggles with the world. This is to say that Christian theology would not be the same, as we know it, if history was different. Once established, however, Christian theology or any theology reaffirms itself by intensifying its position. And the same method is applicable to technology. Although the means are different, the end is similar.

Technology is also developing an evangelicalism that brings the "good news" or gospel of technology to the world. Like many other religions, promoters of technology have an obligation to describe its ideals and to discuss its goals. Neither its importance nor its relevance can be denied. The "good news" about technology is discussed wherever technology exerts its influence, from technical conventions demonstrating the latest widgets and gadgets to hospital operating rooms where life-saving surgery is performed, from landing on the moon to landing on Mars. Technology may be projected as the way to salvation and used for the improvement of everything.

If we find that religions in general no longer embody the qualities we gave them, we may either change those qualities, or create new religions. We may also have no religions at all and live without any anthropomorphisms. If we achieve the latter condition, we may have gone beyond the influence of the axial period discussed above. But religions always have an emotional appeal. A likely possibility is to combine technology with religious rituals and symbols. The result would be a new religion that would be sustained by a theology that is derived from technology's multifaceted applications in the world.

NOTES

1. Arthur Schopenhauer, *Parerga and Paralipomena: Short Philosophical Essays*, trans. E.F.J. Payne, 2 vols., Oxford: Clarendon Press, 1974, II, p. 325.

2. See Ludwig Feuerbach, *The Essence of Christianity*, trans. George Eliot [Marian Evans], 2nd Ger. ed., London: Trübner & Co., 1854, rpt. New York: Harper & Row, 1957, p. 12.

3. Karl Jaspers, *The Origin and Goal of History*, trans. Michael Bullock, New Haven: Yale University Press, 1953, pp. 1-4. Also see John Hick, *An Interpretation of Religion: Human Resources to the Transcendent*, Basingstoke: Macmillan, 1989, pp. 29-31.

4. See chapter four above. It is well-known that the development of technology reveals religious influence that for Western civilization means Christianity, but this knowledge has not encouraged every critic of technology to grant the possibility that technology may evolve into a religion. A religious influence may be granted

to technology, at least historically, but denied to it theologically. Those who study technology and yet deny the possibility of its religion include David F. Noble, *The Religion of Technology: The Divinity of Man and the Spirit of Invention*, New York: Knopf, 1997.

5. Notably, Martin Heidegger, "Overcoming Metaphysics," in his *The End of Philosophy*, trans. Janet Stambaugh, New York: Harper & Row, 1973, pp. 86-87, 93 and 109-110.

6. See Hans Küng, *Theology for the Third Millennium: An Ecumenical View*, trans. Peter Heinegg, New York: Doubleday, 1988, pp. 162-164.

7. *Republic*, II, 379a-III, 392a in *The Collected Dialaogues of Plato, including the Letters*, ed. Edith Hamilton and Huntington Cairns, Princeton: Princeton University Press, 1963, pp. 625-637.

8. *Metaphysics*, VI, 1026a 19 and 1064b 1-2 in *The Complete Works of Aristotle: The Revised Oxford Translation*, ed. Jonathan Barnes, 2 vols., Princeton: Princeton University Press, 1984, II, pp. 1620 and 1681.

9. Far too often, technological theology has been rewritten as an extension of Christianity, or simply denied because Christianity is assumed to be the ultimate source of all truth. Both of these ideas are unrealistic and unsubstantiated. Typical of those who deny that technology has a theology is Guido Schellmann, *Theologie und Technik: 40 Jahre Diskussion um die Technik, zugleich ein Beitrag zu einer Theologie der Technik*, Cologne: Hanstein, 1974, p. 263, who defines all theologies within a Christian context.

10. See *Lessings Werke: Vollständige Ausgabe in fünfundzwanzig Teilen*, eds. Julius Petersen and Waldemar von Olshausen, *et al.*, 25 vols. in 20, Berlin: Deutsches Verlagshaus Bong & Co, 1925, XXIII, pp. 45-50: "Über den Beweis des Geistes und der Kraft." Rendered into English as "On the Proof of the Spirit and of Power," in *Gotthold Ephraim Lessing : Philosophical and Theological Writings*, trans. and ed. H.B. Nisbet, Cambridge: Cambridge University Press, 2005, pp. 83-88. Lessing is famous for demanding proof for the continuation of religion because if a religion is to remain true, it must continue to disclose revelations indefinitely into the future in order to make that religion believable. This idea is discussed and challenged by Soren Kierkegaard, *Concluding Unscientific Postscript to Philosophical Fragments*, ed. and trans. with Introduction and Notes by Howard V. Hong and Edna H. Hong, 2 vols., Princeton: Princeton University Press, 1992, I, pp. 95-106.

11. David Hume, *An Enquiry Concerning Human Understanding*, sect. 10, marginal sect. 90-91 in *Enquiries Concerning Human Understanding and Concerning the Principles of Morals*, ed. L.A. Selby-Bigge, 3rd ed., with text revived and notes by P.H. Nidditch, Oxford: Clarendon Press, 1975, pp. 114-116 and 126-127. Also see Terence Penelhum,"Hume's Criticisms of Natural Theology," in *In Defense of Natural Theology: A Post-Humean Assessment*, eds. James F. Sennett and Douglas Groothuis, Downers Grove (Ill.): InterVarsity Press, 2005, pp. 23-25 for a Christian criticism of Hume's skepticism. Similarly, see Immanuel Kant, *Critique of Pure Reason*, trans. and ed. Paul Guyer and Allen W. Wood, Cambridge: Cambridge University Press, 1998, pp. 586-587 (A636/B664-A638/B666).

Chapter Six

The Future of Religion and the Religion of the Future

INTRODUCTION

Theology predicts the religious environment of the future because it is derived from the religious environment of the past. Different theologies demonstrate this truth because they are grounded within the historical context of their respective religions. Because this relationship is applicable to technology, we can also say that technological theology predicts the religious environment of the future. It challenges the premise on which existing religions are based, and questions their transcendentalism, including godless religions, such as Buddhism and Jainism. It is predicated on the realization of future activities. Because technology is as much about the future as it is about the past, it is as much a forecast of what will occur as an affirmation of what has occurred. Although we do not live in a theocratic age, we do live in an age that anticipates new religious forms. Whether or not they will develop is dependent on the importance we assign to religion, and the choices we will make thereto.

If, for the sake of argument, we were to make a comparison to politics, we might refer to the concept of popular sovereignty, an idea based on the assumption that a few people are empowered to act on behalf of everyone else. Likewise in religion, a person may say that he has been empowered to communicate the wishes of the gods, or that indeed he is a god. Like the concept of popular sovereignty, religion requires the suspension of disbelief. Both are based on a fiction. How close that fiction comes to what is believed to be reality gives it validation. When reality no longer validates the fiction, it becomes exposed to its own absurdity and looses the suspension of disbelief. Just as it is said that governments require consent from the governed, religion requires consent from its believers, although believers have usurped the credit that belongs elsewhere. For example, we give thanks to the gods for saving someone from a life-threatening disease, when in fact, we should give thanks to medical science, regardless of the age we are referring to, for developing medication or surgery that is capable of saving the person.

We also attribute all types of marvels to the blessings of the gods, when in fact credit belongs to humanity and to its achievements. It is humans who study the natural world and devise experimentation and analysis that we call science. It is humans who create all types of methods, procedures, and organizations for life's well-being that we call technology. It is humans who create a way of life subject to laws that we call civil society. It is humans who produce rules of behavior that ethically evolved into what we call morality.

And it is humans who invent religions. As far as gods are concerned, if there are gods, then there must be believers. If there are no believers, then there are no gods, a truth that is applicable to all religions. The Egyptian god Ra, the Toltec god Quetalcoatl, the Greek god Zeus, and the Norse god Odin are not gods, but they were gods. If they are revived, they will be gods again. Therefore, believers are not dependent on gods as much as gods are dependent on believers. A god without believers is no god. Yahweh, Brahman, Jesus, and Allah did not exist as gods before there were people who believed them to be gods. It is people such as Jews, Hindus, Christians, and Muslims who create religions that give them an expression for their beliefs, as knowledge of history will attest. The beginning of a religion is started by persons who are receptive to divine presence, such as Abraham and Muhammad. To say that gods created the world before there were humans to create them is nonsensical. And this condition is true in addition to the view that the creation of the world by one group of people contradicts the creation of the world by another, just as the importance of the Babylonian god Marduk conflicts with the view of the Hebrew god Yahweh as both conflict with the view of Uranus, the earliest supreme god of the Greeks. To say that a god created the world before there were humans to design a religion that contains this idea is an example, to use a cliché, of putting the cart before the horse.[1]

These arguments should not distract us from concluding that technology can become a meaningful religion. If a religion is an outward expression of an inner impression that exemplifies a self-reflective social mechanism in which humanity imposes an interpretation on the world, then technology exemplifies the future. It is removed from the limited worldview of transcendentalist religions. When we realize that a religion is a human creation with characteristics that we desire, then we can create other religions that can assist us without offering temptations of a hypothetical realm that are unrealistic and not helpful to our well-being. What good can be served, for example, when a parent merely prays for a sick and diabetic child to achieve full recovery rather than bringing that child to the hospital for emergency medical treatment? If we know of the benefits of science and technology, it would be foolish not to apply them, but, of course, it is assumed that we will apply them rationally. We are not referring to the use of technologies utilized for destructive purposes, such as in war.

Technology manipulates the world through the accumulation of its effects that includes tools, machines, devices, techniques, methods, and all manner of organization. Since technology is the result of an accumulation that has an influence on world cultures, it may seem as an inevitability that technology can evolve into a religion, even when we consider that the concept of God or god is a mental construct that is subsequently imposed on the world. And it is this ubiquity that intensifies technology's importance.

We must also reiterate what we referred to above regarding revelations: if there are divine beings and they wish to communicate with us, then revelations should be a frequent activity, occurring on a daily, weekly, or monthly basis in order to be valid. A revelation that occurs only once is hardly a validation for anything. Revelations need repetition in order to legitimize their message; otherwise, they would demonstrate abnormality, not divinity. One would think that divine beings would want to communicate with us as often as possible to assure a continuation of our obedience.

As important as anthropomorphism is to some religions, it is not mandatory. There are also some religions without anthropomorphism despite associating divinity with human characteristics. An anthropomorphized religion is Christianity, and an non-anthropomorphized religion is Shintoism. A religion without anthropomorphism that has a god with some human characteristics, but not a human body, is Judaism. Technology as a religion would be similar to Shintoism, but there would be no need to anthropomorphize technology as a religious form because technology already is integrated into every human activity, either by device or method, and assists in human understanding. It is this characteristic of universality that is another reason why technology is a valid candidate for a religion.

In all honesty, we must ask how technology can qualify as a religion if a metaphysical change has not occurred. We have already discussed the development of secularization that first appeared in the thirteenth century, a development that began with contradictions within the Catholic Church, and that helped to establish a foundation to the modern age.[2] But more than secularization, Christian culture itself exerts a strong influence on the development of technology that originated in Western Europe and that later had an effect on the world through westernization. Whether directly relevant to Western civilization or indirectly related to it, technology is now empowered throughout the world with a new paradigm that replaces the old. And it is within this turn of events that a new metaphysical interpretation of the world ensues. No more is technology subordinate to human reality; rather, it has become the most important aspect of it. Technology is applicable everywhere, even in those areas where it was unthinkable to utter its name, particularly in all types of procedures and organizations. As a result, the world has become technologized. Even those who pride themselves in their opposition to technology still use it to make their point.

Nevertheless, we are not deliberately looking for patterns that could led to a demonstrative explanation of religion, although there is a tendency among humans to envision all types of hidden meanings that do not exist, such as demonstrated by the Rorschach inkblot test in which interpretations of shapes are analyzed as indications of intellectual and emotional traits that are thought to be reflections of personality, or predictions for the end of the world. Regardless of our search for meaning, a coincidence is not proof that a tendency, pattern, or repetitious phenomenon exists. Religions originate because of certain circumstances, some of which are easier to explain that others, but they are derived from choices made by people within the human condition. The relevancy of religion is not and has never been applicable universally. Religions are historical manifestations of free choices defined by time and place. The need for religion might be universally applicable, but individual religious expressions of this need are not. The truth of this observation is analogous to technology's presence, not because technology shares the same tendency, but because it has characteristics that could be used to establish itself as a religion. And the establishment of technology as a religion crosses the threshold into unknown territory. Since technology results from adaptations to changing situations because adaptation in general is a characteristic of culture and cultural behavior, it is also a characteristic of technology.

DOMINATION BY TECHNOLOGY

Because we have made ourselves dependent on every manifestation of technology, we have made it our modus operandi. Although we can choose to use less technology, in all likelihood we will use more. In addition to being our handmaiden, technology has also intensified its importance to us. The use of technology has allowed us to combine the human body with mechanical or electrical apparatus that corrects a birth defect, or replaces a body part through prosthesis that was lost by an accident or violence, or controls or aids physiological functions. Technology has also utilized the messages in the brain to help physically impaired or paraplegic people to move or communicate. We have appended machines to humans as we have linked artificial intelligence with human intelligence. Robots of various designs and for distinct purposes have been invented and utilized, and probably will become more commonplace in the future. Because we are instigators of change, the time has passed when we might be free from our own devices.

If the use of technology leads to domination, then it can influence anything, that is, if anything can be influenced by technology, then religions will also succumb. This is not to say that various religions, regardless where they originate, will succumb to technology deliberately, but that they will be influenced by technology as a matter of course because technology is evident everywhere. The importance of this idea brings us full circle to the notion that we live in an age that anticipates new religious forms because technology helps to develop situations that set in place cravings for new expressions. These cravings have a bearing on the perception of religion and the conception of God or gods. For example, at one time, the sun was considered to be so important to many people that it was designated as a god. At other times, iterant preachers were considered to be holy men, or prophets, or even gods. Regardless of their origin, gods are reflective of their age, and this reflection is why they are considered to be gods. In the European Middle Ages, relativism and any connection it may have had with religion would be considered to be blasphemy. Now, it is truth. And truth (*aletheia*), as discussed by the Greeks, means the revelation of what previously was hidden or unknown. The presence of technology is associated with its ubiquity, and its ubiquity is a characteristic of its religiosity, a characteristic that the present age perpetuates.

Since technology is dependent on a method that functions as a framework to which technological phenomena are appended, the combined effect of these phenomena confirms the presence of technology that validates its method, but this method changes whenever technology changes. It is not fixed, but is grounded in the procedure associated with technology that results from technology's accumulative presence. That is to say that technology is derived from its engagement with the world, which is always changing. Although other religions have their own methods, technology's method exerts an influence that few other human endeavors can achieve because it is variable. Its engagement is a confirmation of its applications that deals with change. History without change is ahistorical, and technology without change is atechnological, to coin a term. Technology's application is a demonstration of its facticity, and its facticity establishes a foundation to its religiosity.

What does domination by technology mean? According to Heidegger, technology will be the final phase of being because humanity would have completely sublimated itself to technology,[3] resulting in a realm of artificiality. Even alienation from nature and withdrawal from it would be the same thing. If we agree with the interpretation that technology leads to domination, it does not mean that humanity is enslaved to technology, but that technology has knowingly been conferred with an irreplaceable value because of its benefit to humanity. Although we should be grateful to technology because it is beneficial, this advantage does not mean that we should be oblivious to its influence, or blind to the need to consider alternate possibilities.

Among technologists and technicians, the description "the technological imperative" concerns the idea that if we can do something technically, then we ought to do it, and if we ought to do it, then we must do it. Thus, technology becomes a compulsion through which we exercise our freedom and by which we negate it. It is ironic that although technology emphasizes control over everything, we are losing control over ourselves. Not only the use of robots, but also "virtual reality" symbolize humanity's fall from reality. We live within the artificial world of cyberspace that to some participants seems to be more real, virtually, than reality. Domination by technology ultimately means dehumanization.

Because technology influences how we express ourselves in the world, it helps to create possibilities, and these possibilities exist in proportion to the use we make of technology. Actions by means of technology become the predominant mechanism, among others, of free choice, and this predominance seems to confirm the idea that technology is the only possibility. Hence, when problems exist, technology is thought to be the only solution. As we have mentioned elsewhere, no condition is a determinant of action. It is human beings who determine how technology is applied, although technology is only one of several conditions through which humans act. The domination of technology acquires meaning because when we act on anything, technology is considered to be the only possibility. Once again, technology becomes a compulsion.

As has been described many times before, a discussion of technology ultimately entails a discussion of metaphysics because technology is only one manner by which humans impact reality. Technology benefits us because of its applications, which increases its importance. The metaphysics of technology is linked to the use of ontological freedom because our openness to being is the basis of free choice. What characterizes this openness is a lack within our being, but we are not consumed or defined by it. Technology and its manifestations, including religion, acquire meaning because of free choice. Although technology may become an obstacle to ontological freedom, since it may restrict free choice when other possibilities are not considered, it also acquires meaning because of free choice. Paradoxically, it is simultaneously limited and unlimited by freedom.[4]

THE POTENTIALITY OF TECHNOLOGY AS A RELIGION

In addition to the characteristics that may help technology evolve into a religion, an understanding of religion itself is also pertinent. If religion imposes an interpretation on the world, then technology is a manifestation of a fundamental truth, a truth that technology itself is an interpretation. Needless to say, technology's interpretation is an idea that connects it to ontological freedom. Although not innate to our being as human beings, technology is essential to our existence. We must ask how something that is fundamental to our existence is not also fundamental to our being. Is there a contradiction inherent in this idea? Is the application of our being by means of technology, as influenced by the nearly endless twists and turns technology has manifested over time, related to an essence that is tied to the openness within our being, an openness consumed by ontological freedom? If technology is a result of the exercise of this freedom, then its role is parallel to the exercise of freedom in other endeavors. It is the result of free choice by free beings, that is, it is an example of creativity that originates within us. Of course, technology would have no meaning unless the world existed.

Similarly, a religion must be given an actuality in order to qualify as a religion. Because technology is defined by its engagement with the world, other endeavors are similarly disposed, but technology is self-reinforcing because its essence is the embodiment of engagement. Since technology is expressed through a method that serves as an artifice, it is an excellent candidate for a religion. Obviously, religions demonstrate their own methods that concern the organization and procedures supported by their respective beliefs, as the Catholic Church has a process or formula for those who qualify for sanctification, but technology also contains a method that makes it eminently prepared for success. When applied as a religion, technology is confirmed twofold. It is confirmed by its ontological foundation grounded within the upsurge of our freedom, and it is also confirmed by the characteristics that help establish technology as a religious phenomenon, characteristics that are manifestations of its application, such as its doctrines, rituals, and beliefs, all of which are historically explicable. Thus, technology as a religion reveals its potentiality metaphysically and historically. Although this truth may be denied, its reality cannot be ignored.

As already discussed, new conditions developed in Western Europe that enabled technology to become a serious challenge to Christianity, particularly, when Christian metaphysics was questioned by secularization and when the medieval worldview was thrown into doubt by the Scientific Revolution. This challenge was not caused by direct confrontation, but by pressure exerted gradually, although consistently, on Christianity over a considerable span of time from the sixteenth century to the present. Although slow, this challenge was persistent. After the initial occurrence, secularization and the worldview perpetuated by the new science spread worldwide and influenced other religions and their respective cultures. It is this condition that the world manifests today. Even in Islamic countries, the resistance directed against the West is merely a reaction to the presence of Western values found there, but the growth of religion in many parts of the world does not confirm religious revivalism. It confirms reactions to foreign influence, although religion is used just as often to unite a people politically or culturally as it is used to unite them religiously. It is the means to social interaction.

Because of Christianity, Western technology is characterized by advancement on the one hand and mediocrity on the other. Technology's advancement is influenced by the Christian idea of improvement that originated on the personal level, while mediocrity is the result of Christianity's intention of reaching and hopefully saving everyone. Advancement in Christianity is associated with the idea of progress—whether or not it is historically verifiable—that promotes a linear conception of time for the attainment of one purpose, the salvation of souls and their enjoyment with God in heaven. Its influence on technology is pointedly progressive, characteristically linear, and intended to introduce improvement wherever possible. These attributes are associated with Christianity and the Christianized type of technology that were perfected in Western Europe. Although seeming to be contradictory to advancement, mediocrity is the practical effect of this achievement, both for Christianity and technology.

Despite this development, we must ask if the concepts that apply to religion are also applicable to technology. Knowing that technology is an artificiality that results from our creativity, we are not referring to the characteristics of religion (sociability, practicality, and a belief system), or to the historical influences to modern (originally, Western) technology as it evolved through the Middle Ages, or to the process of Christianization relevant to Western technology, but to the choices made by free beings. Apart from the idea associated with technology as a type of applied conceptualization, this idea conflicts with the fact that technology is supplemental to human evolution. It has not been determined that technology is derived from anything other than free choice. It is the result of humanity's presence in the world because the world is a challenge to existence. Technology originates with the satisfaction of basic needs that are conditioned by circumstance. If humans were not free to create technology, it would not exist, but humanity also would not exist because technology is essential to human survival. Although technology is not innate to our being, it is essential to our existence.

Nor is technology removed from human experience. Because it is tied to the world as the means by which it is applied, its applications reinforce practicality. What does technology's practicality have in common with Christianity's practicality? And how does technology relate to mediocrity, that is, to the practical effect of Christianity's influence? Being receptive to invention, and especially to innovation, technology has become highly adaptive. Adaptation allows technology to exert itself, which means of course, that we allow technology to influence our choices. If one devise, or method, or machine has little effect on an individual, other devices, or methods, or machines could. Therefore, the goal of technology, unless it is explicitly perfected in isolation, like the ancient Chinese silk industry, is to influence everyone, and what influences everyone is the lowest common denominator, which is the most commonly shared trait. When the latter occurs, mediocrity appears, spreading out to the world at large. Not only is practicality a characteristic of religion, but it is also a characteristic of technology's application.

CONCLUSION

The question we have before us concerns whether or not technology can successfully challenge religion and replace it with itself. Given that technology has been influenced by the changes occurring within European society, especially during the Middle Ages when Christianity attained political and cultural superiority, it should be supposed that technology would adopt, in modified form, Christian attributes. These attributes concern Christianity's forwardly directed belief system that regards itself to be the culmination of human existence. Because it begs the question, Christianity creates a cause (original sin) in order to justify its necessity (the introduction of divine grace that obliterates original sin). Christianity is based on an elemental interpretation of suffering within the context of human existence, while creating a transcendental realm in which every event will be judged by a divine rubric. This transcendental realm is perceived to be the apogee of reality, while the world is demoted to its nadir. In order to achieve its goals, Christianity emphasizes means over ends. If the end is eternal salvation, then the means are of vital importance. More than any specific means employed, it is the idea of means itself that is emphasized. Since technology assumes these attributes, it has become forwardly directed, emphasizing means over ends, offering itself as the culmination of human reality, except that its reality is worldly, not otherworldly.

Technology has also adopted from Christianity the idea that only an external entity can be knowledgeable about truth. Supporters of Christian doctrine say that only Christianity is able to define what is right and wrong. Similarly, when this idea is applied to technology, it follows that only technology is capable of defining truth. Technology becomes the criterion for a new paradigm, the standard of a new beginning.

Knowledge of the history of technology in the West and of technology's westernization in non-Western cultures has enabled us to be aware of two related possibilities. The first possibility concerns the acceptance of this new paradigm. It concerns the substitution of technology for religion that would continue the latter, but it would be within a technological framework, including its effects on economics and politics. The second possibility concerns the opposite condition in which technology as a potential religion is rejected, a rejection of which would intensify nihilism, which is itself the effect of the Scientific Revolution and the dissolution of Christian (or Platonic-Christian) metaphysics. Nihilism would not only destroy itself, but it would also destroy everything else. Life would be meaningless, since there would be neither hope for a transcendental realm, nor hope for an earthly one. Because religion promotes morality, the absence of the latter would make existence amoral, completely devalued of meaning. If nihilism is to be overcome, a new conception of reality must be created. The way in which humanity can be preserved is by means of technology, so the latter would be the logical choice for preservation. Of these two possibilities, the first possibility is more hopeful and more likely.

The context of technology's enduring presence is related to the idea that its creation is the result of humanity's struggle for existence. This idea is apart from and predates Christianity's idea that it is the explanation for human suffering. This struggle signifies more than physical survival. It also signifies the manner by which humans achieve the fulfillment of their capabilities. Therefore, it denotes self-realization in the most complete sense. Because it relates to the fulfillment of human existence, it is an ontological description, dealing with improvement, growth, and development. If this struggle is applicable individually, can it also apply to society? Can it apply to the mutual interests and shared institutions of social organization? Knowing that society integrates individuals into groups, the purpose of integration concerns cooperation among its members. The values and morals of its members reinforce society's organization, which in turn maintains its culture. Hence, society is made real through the exercise of this struggle. Within its dynamics, society is born. Influenced by historical and cultural choices, society is shaped by the human condition that lays the foundation to human reality, the condition that is characterized by many temporal situations, including suffering, care and concern, inwardness, dread, guilt, and loneliness, among others. All of these situations are shaped by ontological freedom. The human condition lays the foundation through which individuals create everything, including religion. Therefore, religion is an ontological manifestation.

Human reality, like the Nietzschean will to power, is dependent on interpretation, although it is more accurate to say that both are its embodiment, despite what governments, or the church, or science prescribes. An interpretation denotes the application of a conception of some type of reality, whether that reality is based, for example, on historical analysis, scientific experimentation, political maneuvering, or emotional assessment. An interpretation explains something by means of the conditions that bear on it, and for us, these are the conditions of life expressed through history and culture. No interpretation is final because any interpretation is merely an explanation of the world, an approximation that presents an understanding within a specific time and place.

The interpretive manifestation of any reality brings us full circle to our definition of religion above, and this definition is applicable to past, present, or future religions. If religion is an outward expression of an inner impression that exemplifies a self-reflective social mechanism in which humanity imposes an interpretation on the world, then religion is an explanation for human reality. A religion is simply one explanation of the totality of all explanations, which means that any religion offers only one interpretation of the world. Therefore, no religion is closer than any other religion to an ultimate truth. It is true to say that a religion reflects the age and the culture in which it appears. Religions are historical creations, brought into the world by individuals and societies, and shaped by the events of their time. Reciprocally, the religious importance of technology is a reflection of our time. It epitomizes the most important phenomenon in human culture. It is the embodiment of our endeavors, applicable wherever we go, always at hand to alter our choices. Because the past is partially responsible for empowering technology, technology can rely on the past to direct its future.

NOTES

1. See chapter one above that refers to mental objects or formulations that concern human interaction with the world.

2. Nevertheless, religion may survive even if society is secularized, but its survival is dependent on religion's privatization.

3. Martin Heidegger, "Overcoming Metaphysics," in his *The End of Philosophy*, trans. Janet Stambaugh, New York: Harper & Row, 1973, pp. 86-87, 93 and 109-110.

4. In general, see my "An Introduction to the Metaphysics of Technology," *Technology in Society*, 27 (2005), 558-560.

Bibliography

Agamben, Giorgio, *The Signature of All Things: On Method*, trans. Luca D"Isanto with Kevin Attell, New York: Zone Books, 2009.

Allen, Michael J.B., Valery Rees, with Martin Davies, eds., *Marsilio Ficino: His Theology, His Philosophy, His Legacy*, Leiden: Brill, 2002.

Ashbrook, James B., *The Human Mind and the Mind of God: Theological Promise in Brain Research*, Lanham: University Press of America, 1984.

Augustine, *The City of God against the Pagans*, ed. and trans. R.W. Dyson, Cambridge: Cambridge University Press, 1998.

Bennett, Maxwell *et al.*, *Neuroscience and Philosophy: Brain, Mind, and Language. With an Introduction and Conclusion by Daniel Robinson*, New York: Columbia University Press, 2007.

Bennett, M[axwell] R. and P.M.S. Hacker, *Philosophical Foundations of Neuroscience*, Oxford: Blackwell, 2003.

Benz, Ernst, *Evolution and Christian Hope: Man's Concept of the Future from the Early Fathers to Teilhard de Chardin*, trans. Heinz G. Frank, Garden City: Doubleday, 1966.

Bergson, Henri, *The Two Sources of Morality and Religion*, trans. R. Ashley Audra and Cloudesley Brereton, with the assistance of W. Horsfall Carter, New York: Henry Holt, 1935.

Blumenberg, Hans, *The Legitimacy of the Modern Age*, 2nd rev. ed., trans. Robert M. Wallace, Cambridge [Mass.]: MIT Press, 1983.

Boyer, Pascal, *Religion Explained: The human instincts that fashion gods, spirits and ancestors*, London: Heinemann, 2001.

Bradley, F.H., *Appearance and Reality: A Metaphysical Essay*, 2nd ed., Oxford: Clarendon Press, 1897, rpt. 1930.

Bultmann, Rudolf, *History and Eschatology*, Edinburgh: Edinburgh University Press, 1957.

Butler, [Edward] Cuthbert, *Benedictine Monachism: Studies in Benedictine Life and Rule*, 2nd ed., London: Longmans, 1924, rpt. Cambridge: Speculum Historiale, 1961.

———, "Saint Benedict and the Sixth Century," *Downside Review*, 48 (1930), 179–197.

Changeux, Jean-Pierre, *Neuronal Man: The Biology of Mind*, trans. Laurence Garey, New York: Pantheon Books, 1985.

———, *Physiology of Truth: Neuroscience and Human Knowledge*, trans. M.B. DeBevoise, Cambridge [Mass.]: Belknap Press, 2004.

Chapman, Dom John, *Saint Benedict and the Sixth Century*, New York: Longmans, 1929.

Cohn, Norman, *The Pursuit of the Millennium: Revolutionary Millenarians and Mystical Anarchists of the Middle Ages*, 3rd rev. ed., New York: Oxford University Press, 1970.

The Collected Dialogues of Plato, including the Letters, ed. Edith Hamilton and Huntington Cairns, Princeton: Princeton University Press, 1963.

The Complete Works of Aristotle: The Revised Oxford Translation, ed. Jonathan Barnes, 2 vols., Princeton: Princeton University Press, 1984.

Cottingham, John, *et al.*, trans., *The Philosophical Writings of Descartes*, 3 vols., Cambridge: Cambridge University Press, 1985–1991.

Crombie, A.C., *Augustine to Galileo*, 2nd ed., 2 vols., London: Mercury Books, 1961.

Cullmann, Oscar, *Christ and Time: The Primitive Christian Conception of Time and History*, trans. Floyd V. Filson, rev. ed., Philadelphia: Westminster Press, 1964.

The Defensor Pacis of Marsilius of Padua, ed. C.W. Previté-Orton, Cambridge: [Cambridge] University Press, 1928.

de Vogüé, Adalbert and Jean Neufvillle, eds., *La règle de saint Benoît*, 7 vols., Paris: Éditions du Cerf, 1971–1977.

Dilthey, Wilhelm, *Introduction to the Human Sciences*, ed. with an Introduction [and Translation] by Rudolf A. Makkreel and Frithjof Rodi, Princeton: Princeton University Press, 1989.

Durkheim, Emile, *The Elementary Forms of the Religious Life*, trans. Joseph Ward Swain, London: George Allen, 1915.

Eisenstein, Elizabeth L., *The Printing Press as an Agent of Change: Communications and cultural transformations in early Modern Europe*, Volumes I & II, Cambridge: Cambridge University Press, 1980.

Eliade, Mircea, *A History of Religious Ideas*, trans. Willard R. Trask, *et al.*, 3 vols., Chicago: University of Chicago Press, 1978–1985.

Eusebius of Caesarea, *Ecclesiastical History*, trans. Kirsopp Lake and J.E.L. Oulton, 2 vols., Loeb Classical Library, Cambridge [Mass.]: Harvard University Press, 1926–1932.

Fenn, Richard K., *Time Exposure: The Personal Experience of Time in Secular Societies*, New York: Oxford University Press, 2001.

Feuerbach, Ludwig, *The Essence of Christianity*, trans. George Eliot [Marian Evans], 2nd Ger. ed., London: Trübner & Co., 1854, rpt. New York: Harper & Row, 1957.

Frank, Erich, *Philosophical Understanding and Religious Truth*, New York: Oxford University Press, 1945, rpt. 1966.

Geertz, Clifford, "Religion as a Cultural System," in *Anthropological Approaches to the Study of Religion*, ed. Michael Banton, London: Tavistock Publications, 1966, pp. 1–46.

Gimpel, Jean, *The Medieval Machine: The Industrial Revolution of the Middle Ages*, 2nd ed., Aldershot: Wildwood House Ltd., 1988.

Gotthold Ephraim Lessing: Philosophical and Theological Writings, trans. and ed. H.B. Nisbet, Cambridge: Cambridge University Press, 2005.

Grant, Edward, ed. and trans., *Nicole Oresme and the Kinematics of Circular Motion: Tractatus de commensurabilitate vel incommensurabilitate motuum celi*, Madison: University of Wisconsin Press, 1971.

Guthrie, Stewart Elliott, *Faces in the Clouds: A New Theory of Religion*, New York: Oxford University Press, 1993.

Heidegger, Martin, *The End of Philosophy*, trans. Janet Stambaugh, New York: Harper & Row, 1973.

———, *Holzwege*, Frankfurt: Klostermann, 1950.

———, *Off the Beaten Track*, ed. and trans. Julian Young and Kenneth Haynes, Cambridge: Cambridge University Press, 2002.

Hick, John, *An Interpretation of Religion: Human Resources to the Transcendent*, Basingstoke: Macmillan, 1989.

Hicks, G. Dawes, *The Philosophical Bases of Theism*, London: George Allen & Unwin, 1937.

Hume, David, *Enquiries Concerning Human Understanding and Concerning the Principles of Morals*, ed. L.A. Selby-Bigge, 3rd rev. ed. P.H. Nidditch, Oxford: Clarendon Press, 1975.

James, William, *The Varieties of Religious Experience: A Study of Human Nature*, New York: Longmans, Green & Co., 1902.

Jaspers, Karl, *The Origin and Goal of History*, trans. Michael Bullock, New Haven: Yale University Press, 1953.

Jonas, Hans "Seventeenth Century and After: The Meaning of the Scientific and Technological Revolution," in his *Philosophical Essays: From Ancient Creed to Technological Man*, Englewood Cliffs: Prentice Hall, 1974, pp, 45–80.

Kant, Immanuel, *Critique of Pure Reason*, trans. and ed. Paul Guyer and Allen W. Wood, Cambridge: Cambridge University Press, 1998.

Kierkegaard, Soren, *Concluding Unscientific Postscript to Philosophical Fragments*, ed. and trans. with Introduction and Notes by Howard V. Hong and Edna H. Hong, 2 vols., Princeton: Princeton University Press, 1992.

Knowles, David, *Christian Monasticism*, New York: McGraw-Hill, 1969.

Küng, Hans, *Theology for the Third Millennium: An Ecumenical View*, trans. Peter Heinegg, New York: Doubleday, 1988.

Kuhn, Thomas S., *The Copernican Revolution: Planetary Astronomy in the Development of Western Thought*, Cambridge [Mass.]: Harvard University Press, 1957.

Lactantius, *The Divine Institutes, Books I-VII*, trans. Sister Mary Francis McDonald, Washington: Catholic University of America Press, 1964.

Leiss, William, *The Domination of Nature*, New York: Braziller, 1972.

Lessings Werke: Vollständige Ausgabe in fünfundzwanzig Teilen, eds. Julius Petersen and Waldemar von Olshausen, *et al.*, 25 vols. in 20, Berlin: Deutsches Verlagshaus Bong & Co., 1925.

Locke, John, *An Essay Concerning Human Understanding*, ed. Alexander Campbell Fraser, 2 vols., Oxford: Oxford University Press, 1894.

Loewe, Michael, "Cyclical and Linear Concepts of Time in China," in *The Story of Time*, ed. Kristen Lippincott, *et al.*, London: Merrell Holberton, 1999, pp. 76–79.

Löwith, Karl, *Martin Heidegger and European Nihilism*, ed. Richard Wolin, and trans. Gary Steiner, New York: Columbia University Press, 1995.

MacIntyre, Alasdair and Paul Ricoeur, *The Religious Significance of Atheism*, New York: Columbia University Press, 1969.

McTaggart, John McTaggart Ellis, *Some Dogmas of Religion*, 2nd ed., London: Edward Arnold, 1930.

Marshall, Henry Rutgers, *Instinct and Reason: An Essay concerning the Relation of Instinct to Reason, with some Special Study of the Nature of Religion*, New York: Macmillan, 1898.

Marsilius of Padua, The Defender of the Peace, Translation and Introduction by Alan Gewirth, 2 vols., New York: Columbia University Press, 1951–1956.

Martin, David, *On Secularization: Towards a Revised General Theory*, Aldershot: Ashgate Publishing, 2005.

Martineau, James, *A Study of Religion: Its Sources and Contents*, 2nd rev. ed., 2 vols., Oxford: Clarendon Press, 1900.

Morris, Colin, *The Discovery of the Individual, 1050–1200*, London: Society for Promoting Christian Knowledge, 1972, rpt. Toronto: University of Toronto Press, 1987.

Mowinckel, S[igmund], *He That Cometh*, trans. G.W. Anderson, New York: Abingdon Press, n.d. [1954].

Mumford, Lewis, *The Myth of the Machine*, 2 vols., New York: Harcourt, Brace, Jovanovich, 1966–1970.

Niermeyer, J.F., *Mediae Latinitatis Lexicon Minus*, ed. C. van de Kieft, Leiden: Brill, 1976.

Nietzsche, Friedrich, *The Gay Science*, trans. with Commentary by Walter Kaufmann, New York: Vintage Books, 1974.

———, *Gesammelte Werke*, Musarionausgabe, 23 vols., Munich: Musarion Verlag, 1922–1929.

Noble, David F., *The Religion of Technology: The Divinity of Man and the Spirit of Invention*, New York: Knopf, 1997.

Orosius, *The Seven Books of History against the Pagans*, trans. Roy J. Deferrari, Washington: Catholic University of America Press, 1964.

Ortega y Gasset, José, *The Revolt of the Masses*, anonymous English trans., New York: W.W. Norton, 1932.

Otto, Rudolf, *The Idea of the Holy: An Inquiry into the non-rational factor in the idea of the divine and its relation to the rational*, trans. John W. Harvey, 2nd ed., London: Oxford University Press, 1950.

Ovitt, Jr., George, *The Restoration of Perfection: Labor and Technology in Medieval Culture*, New Brunswick: Rutgers University Press, 1987.
Parsons, Talcott, *The Social System*, Glencoe: Free Press, 1951.
Pascal, Blaise, *Pensées* in *Oeuvres complétes*, préface d'Henri Gouhier, présentation et notes de Louis Lafuma, Paris: Éditions du Seuil, 1963.
———, *Pensées*, trans. A.J. Krailsheimer, Harmondsworth: Penguin, 1966.
Passmore, John, *Man's Responsibility for Nature: Ecological Problems and Western Traditions*, New York: Scribner's, 1974.
Patterson, Robert Leet, *A Philosophy of Religion*, Durham: Duke University Press, 1970.
Penelhum, Terence, "Hume's Criticisms of Natural Theology," in *In Defense of Natural Theology: A Post-Humean Assessment*, eds. James F. Sennett and Douglas Groothuis, Downers Grove [Ill.]: InterVarsity Press, 2005, pp. 14–42.
Persinger, Michael A., *Neuropsychological Bases of God Beliefs*, New York: Praeger, 1987.
Pratt, James Bissett, *The Religious Consciousness: A Psychological Study*, New York: Macmillan, 1920.
Rivers, Theodore John, "An Introduction to the Metaphysics of Technology," *Technology in Society*, 27 (2005), 551–574.
———, "The Conception of Time and Its Relationship to Technology," *Research in Philosophy and Technology*, 19 (2000), 215–231.
———, *Contra Technologiam: The Crisis of Value in a Technological Age*, Lanham: University Press of America, 1993.
———, "Technology and Religion: A Metaphysical Challenge," *Technology in Society*, 28 (2006), 517–531.
Santayana, George, *The Life of Reason, or the Phases of Human Progress*, 5 vols., New York: Charles Scribner's Sons, 1905–1906.
Sartre, Jean-Paul, *Being and Nothingness: An Essay on Phenomenological Ontology*, trans. and with an introduction by Hazel E. Barnes, New York: Philosophical Library, 1956.
Schellmann, Guido, *Theologie und Technik: 40 Jahre Diskussion um die Technik, zugleich ein Beitrag zu einer Theologie der Technik*, Cologne: Hanstein, 1974.
Schleiermacher, Friedrich, *The Christian Faith*, 2nd Ger. ed., trans. H.R. Mackintosh and J.S. Stewart, Edinburgh: T.T. Clark, 1928.
Schopenhauer, Arthur, *Parerga and Paralipomena: Short Philosophical Essays*, trans. E.F.J. Payne, 2 vols., Oxford: Clarendon Press, 1974.
Sharpe, Eric J., *Comparative Religion: A History*, 2nd ed., London: Duckworth, 1986.
Smith, Wilfred Cantwell, *The Meaning and End of Religion: A New Approach to the Religious Traditions of Mankind*, New York: Macmillan, 1963.
Stark, Rodney and William Sims Bainbridge, *The Future of Religion: Secularization, Revival and Cult Formation*, Berkeley: University of California Press, 1985.
Supplementum ficinianum. Marsilii Ficini florentini philosophi platonici opuscula inedita et dispersa, ed. Paul Oskar Kristeller, 2 vols., Florence: Leon S. Olschki, 1937.
Tertullian, *Apology*, trans. T.R. Glover, Loeb Classical Library, Cambridge [Mass.]: Harvard University Press, 1931.

Theophilus, *The Various Arts*, trans. with Introduction and Notes by C.R. Dodwell, London: Thomas Nelson & Sons, 1961.

Tschannen, Olivier, *Les théories de la sécularisation*, Geneva: Librairie Droz, 1992.

Vattimo, Gianni, *After Christianity*, trans. Luca D'Isanto, New York: Columbia University Press, 2002.

White, Jr., Lynn, *Medieval Religion and Technology: Collected Essays*, Berkeley: University of California Press, 1978.

———, "What Accelerated Technological Progress in the Western Middle Ages?" in *Scientific Change: Historical studies in the intellectual, social and technical conditions for scientific discovery and technical invention, from antiquity to the present*, ed. A.C. Crombie, New York: Basic Books, 1963, pp. 272–291.

Whitrow, G.J., *Time in History: The Evolution of Our General Awareness of Time and Temporal Perspective*, Oxford: Oxford University Press, 1988.

Wittgenstein, Ludwig, *Tractatus Logico-Philosophicus*, trans. C.K. Ogden, with an Introduction by Bertrand Russell, London: Routledge & Kegan Paul, 1922.

Workman, Herbert B., *The Evolution of the Monastic Ideal from the Earliest Times down to the Coming of the Friars: A Second Chapter in the History of Christian Renunciation*, London: Epworth Press, 1913, rpt. Boston: Beacon Press, 1962.

Index

www.ingramcontent.com/pod-product-compliance
Ingram Content Group UK Ltd.
Pitfield, Milton Keynes, MK11 3LW, UK
UKHW041122071026
15587UKWH00009B/77

9 780761 856542